NEW TOP 50 VILLA

最新别墅设计 50 例 Ⅱ

深圳市海阅通文化传播有限公司　主编

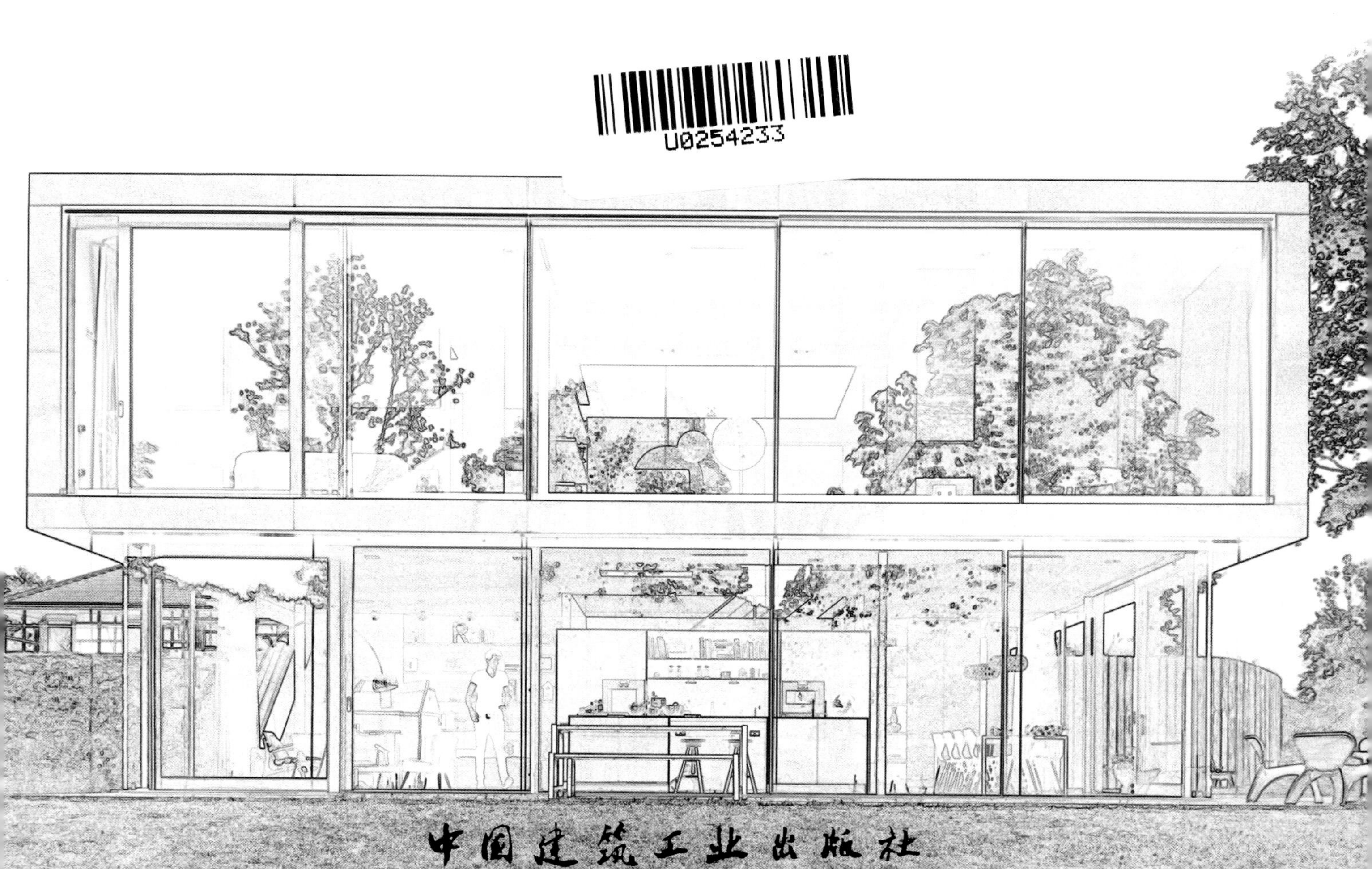
U0254233

中国建筑工业出版社

序言 PREFACE

As graceful as music, the architecture is the embodiment of desirable art. Every architect has the first desire to make the unique building appearance while interior designers expect to provide elegant and comfortable interior design. The outstanding villa design has an appealing but not showy appearance, as well as reasonable spatial scale and creative but not exaggerated color scheme, whilst it uses environmental but special materials. A perfect villa design can provide its owner with identity of noble status and comfort level.

The New Top 50 Villa has gained a good reputation from designers since its publication. This series of books is committed to presenting the latest and the best villa design which is the most stylish and comprehensive. 50 latest projects from more than a dozen countries like America, Brazil, Singapore, are included. Moreover, it provides comprehensive interpretation and appreciation for building appearance and house furnishings, etc. The New Top 50 Villa II inherits the spirit and concepts of the previous book. It requires a higher level of the project source, complete time, area, style of both the architecture and interior design, whilst it will display every project's materials and design highlights from styles to colors.

——by Editorial Committee

建筑是流动的音符，是优美的乐章，是让人赏心悦目的艺术的具体呈现。建筑外观别致、新颖；室内陈设优雅、舒适是每一位建筑设计师和室内设计师在设计之初的愿景。优秀的别墅设计，具有醒目但不张扬的建筑外观、环保但不平庸的设计材料、合理但不单调的空间尺度以及跳跃但不夸张的色彩搭配等。完美的别墅可以使每一位入住的业主对身份尊贵感和生活舒适度产生认同。

《最新别墅设计 50 例》自出版以来，受到许多设计师朋友们的喜爱和追捧。本系列图书立足于为读者朋友们呈现最新、最全、最具风格和特点的别墅设计，收纳来自美国、新加坡、巴西等在内的十来个国家的 50 个最新设计案例，从别墅外观、室内陈设等多方面全方位地解读和分享。《最新别墅设计 50 例 II》延续上一本的精神和立意，在项目来源、项目面积、项目建筑风格和室内设计风格和项目完成时间等多方面有了更高的要求，从外观到室内，从宏观风格到微观色彩和材料的展示，更加细致和完善。

——编辑委员会

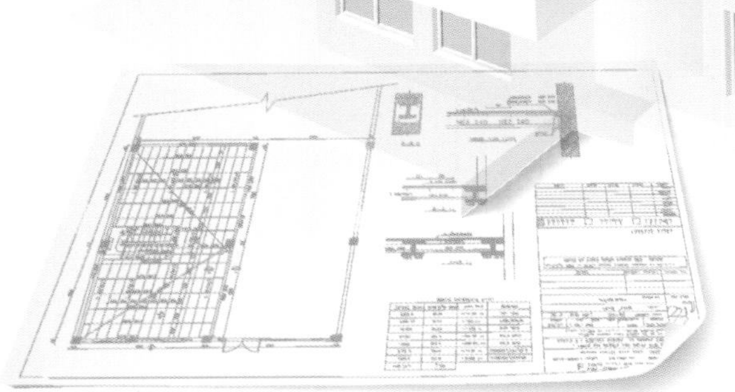

目录

CONTENTS

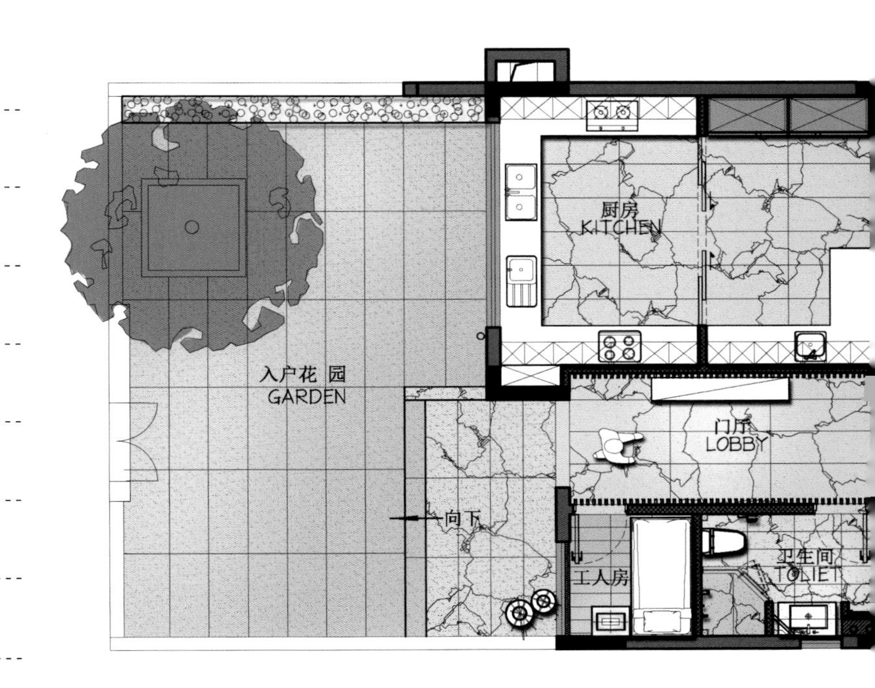

目录
CONTENTS

目录

CONTENTS

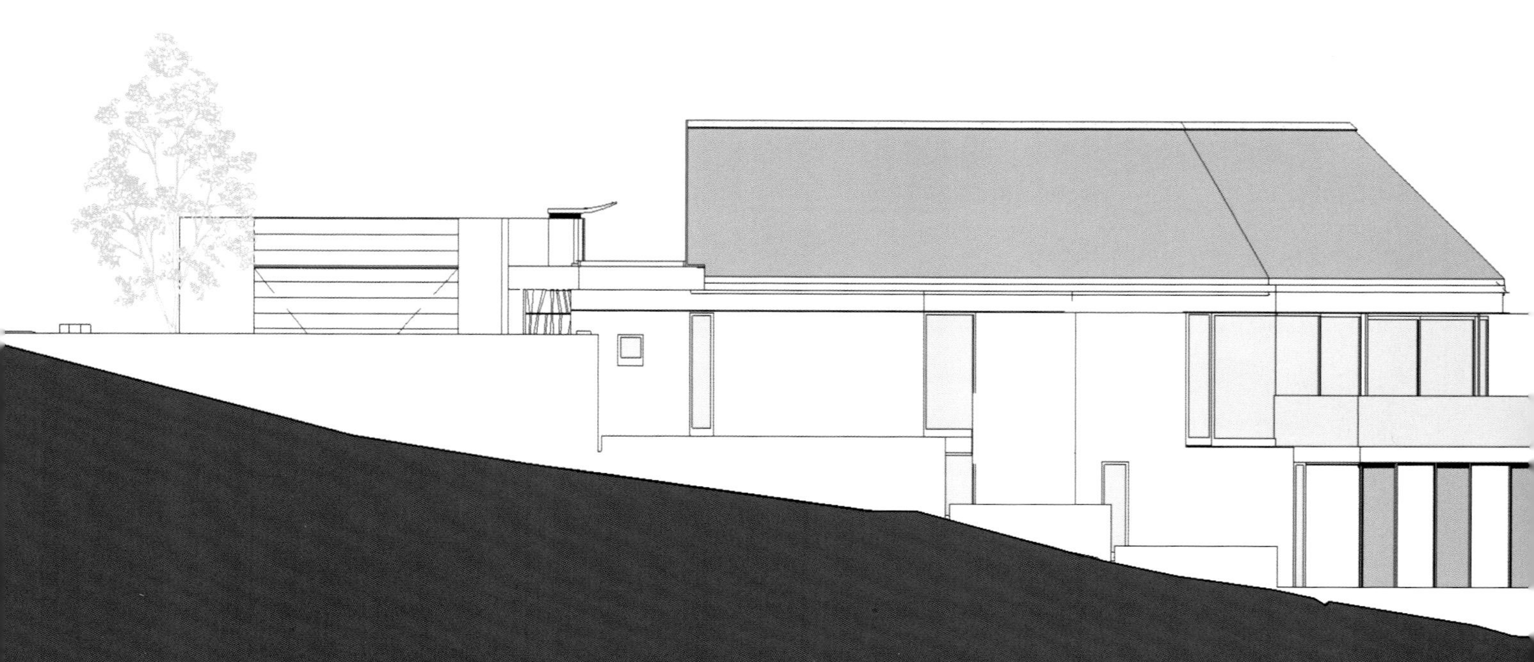

Villa of Golden Sunshine City

阳光金城别墅

- Design Agency: Pinki Interior Design Liu& Associates IARI Interior Design
- Designer: Danfu Liu
- Location: Xi' an, Shanxi
- Area: 600m^2

- 设计单位：PINKI（品伊国际创意）& PINKI DESIGN 美国 IARI 刘卫军设计事务所
- 设计师：刘卫军
- 项目地点：陕西西安
- 项目面积：600m^2

This project is designed as a personal club with the concept of a reception place or an exhibition hall. The hallway including a shoes changing room, leads to the living room and an extra room. Three panoramic windows have a total length of 29 meters. The pond between the two rooms is designed to separate and connect the two spaces. The themed wall decorated with culture stones beside the staircase creates a more friendly conversation atmosphere. The dining area, by the side of a big kitchen of 22 square meters, includes a main dining room and a breakfast room. Besides, a gallery, a water bar, a family cinema and an area for afternoon tea are arranged well in the first floor.

To go up into the personal living space, you can take the lift beside the dining room or the stair connecting the central garden and the house. The style of the space integrates Western into Chinese style. Two rooms on the second floor have the independent coatroom and the washroom. A large terrace shared by the rooms has a good view of the garden. The public area is divided into two parts, the main family room for family members and the extra room for friends. Through the family room, we can get into the master's bedroom with multiple functions and it has a terrace both in and out of the room. The stair beside it leads to the private space for writing and relaxing. With the same design style of the mountain home with the first floor, it has a study, a yoga room, a washroom and a terrace.

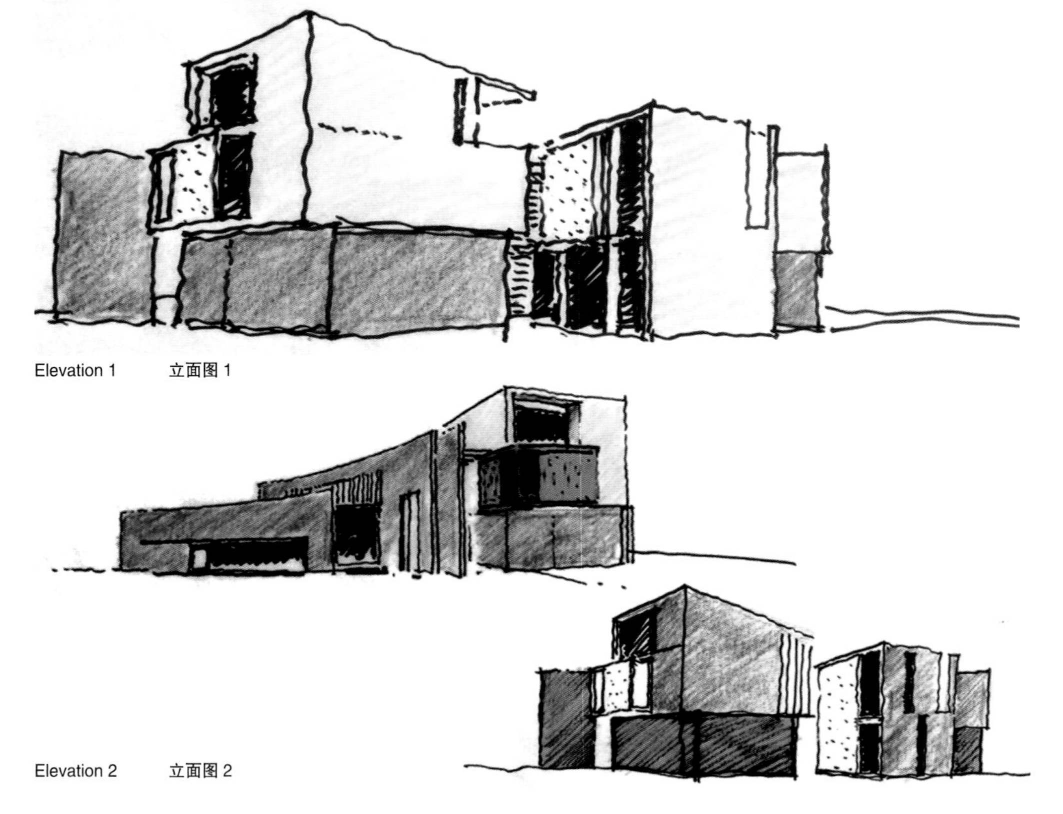

Elevation 1　立面图 1

Elevation 2　立面图 2

该私人会所的设计接近于招待会所或艺术展厅的概念。进入会所之后，经过配有换鞋间功能的门厅，首先来到以洽谈为主的副厅及客厅。三面共约 29m 长的落地窗的环景视野，铺陈副厅及客厅的宽阔尺度，两厅之间的景观及下沉水池起功能分区及过渡之用，利用手扶梯位置设计了文化石铺设的主题墙，让副厅及客厅的洽谈范围更具亲和力。通过主题墙的门洞即是早餐厅及餐厅，餐厅旁边配有 $22m^2$ 的大厨房。整体而言，一层私人会所区域包括一道围绕极佳采光的中心庭院廊道、水吧区、早餐厅、主餐厅、家庭影院、午茶区（木平台）等，从传统大而无当的制式排场概念跳脱出来。

进入私密起居空间，可通过与中心庭院相连的楼梯和餐厅旁边的升降电梯两种方式。居住空间注入了中西混搭风，两间次卧都配有独立的衣帽间及卫生间，透过两间次卧共有的大露台可观赏园区景色。在公共区域分别设有供家庭成员休息交流的主家庭厅及供主人休息或接待密友的副家庭厅。通过家庭厅可进入功能配套齐全的主卧，同时还拥有内外露台。主卧小门旁的楼梯可通往主人三层私密性休闲及写作空间，主要延续了一层追崇自然的山居设计风格，配有洗手间、书房、瑜伽室、内露台、外露台。

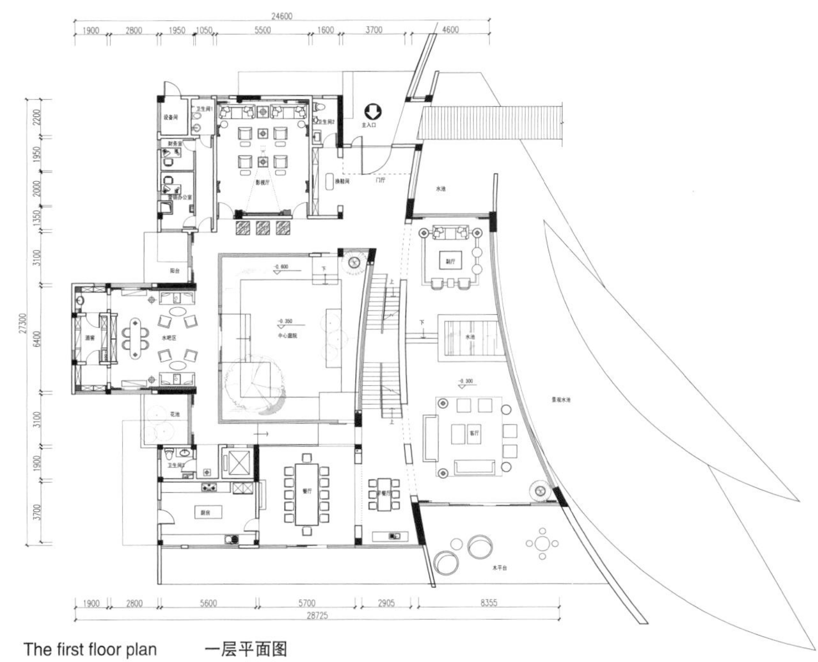

The first floor plan　　一层平面图

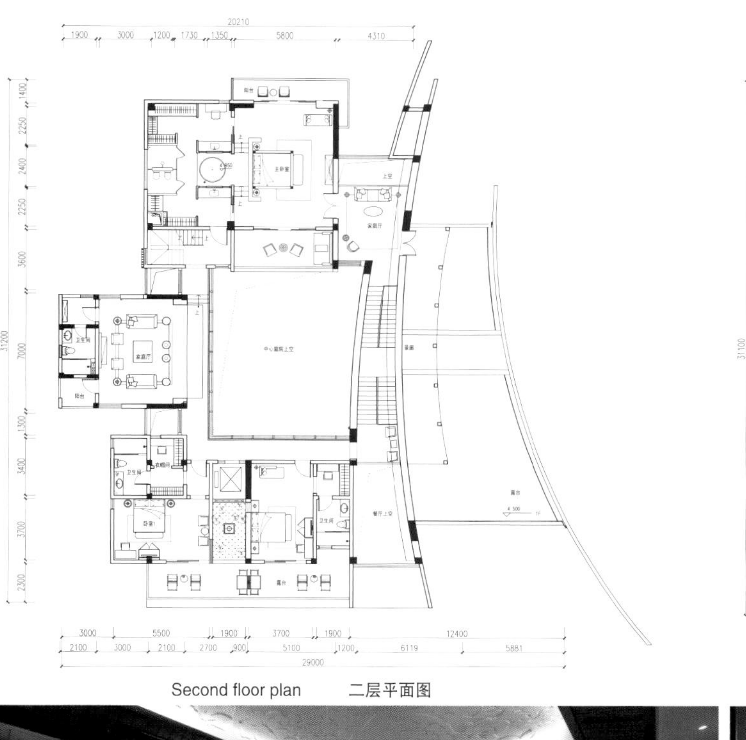

Second floor plan 二层平面图

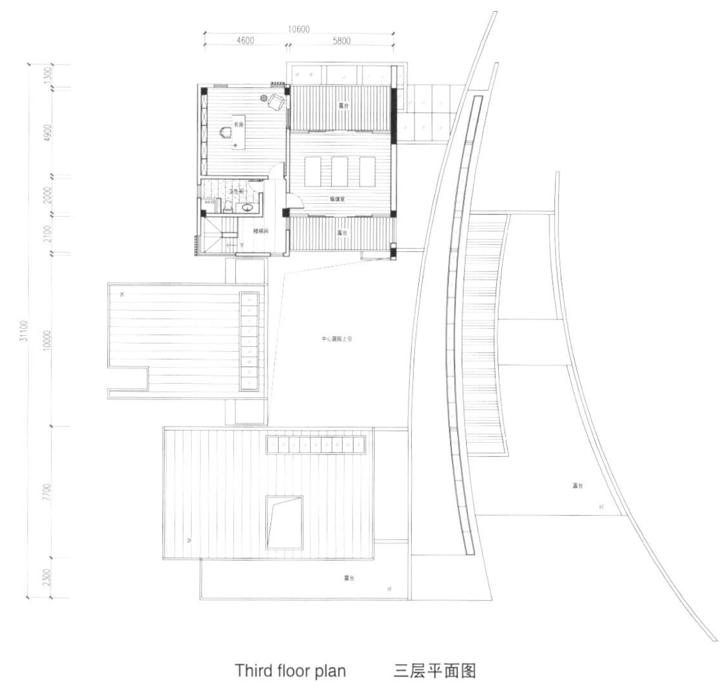

Third floor plan 三层平面图

Heritage and Modernity

传统魅力的现代别墅

- Design Agency: Luigi Rosselli Architects
- Location: Sydney, Australia
- Photographer: Justin Alexander & Edward Birch

- 设计单位：Luigi Rosselli 建筑设计机构
- 项目地点：澳大利亚，悉尼
- 摄影：贾斯汀・亚历山大，爱德华・伯奇

This Woollahra house weaves the rich heritage character of an original Victorian house, with elements that are unashamedly contemporary. Stripped of years of accretions, the original house has been exposed and brought to the fore. A new spiralling stair knits a capacious modern wing, including a new kitchen and family room on the ground floor and study and staff quarters above, with the historic fabric of the dwelling. Steel windows engineered to their uttermost extent fully open the family room onto an expansive outdoor terrace dappled in light. The interiors have been white washed to emphasis the grandeur of the existing house, and draw attention to the dense clutch of trees that surrounds the garden and frames views of the harbour. Wide oak floorboards and finely veined marble lend a timeless air. The simplicity of palette and detail within the house is offset by dramatic detailed light installation by Lindsey Adelman. This grand old residence is not just class though, she also has a conscience. At the instructions of the client, the project incorporates many energy-conserving and generating technologies including our first installation of geothermal air-conditioning/heating system.

本案是一栋富有传统魅力的维多利亚式建筑，整体的设计还大胆加入了现代元素。除去时间在这栋建筑上留下的痕迹，建筑本身的样子便得到凸显。新安装的梯子盘旋而上，连接各个空间，包括一楼的厨房和家庭活动室、二楼的书房和房间，同时保留了住宅的历史结构。钢结构玻璃窗的设计是为了最大化地实现室内外空间的沟通，使室内空间可以向身处斑驳光影中的户外露台敞开。为了凸显别墅的庄严，室内的墙壁被统一刷成白色，让身处其间的人自然而然地将视线投向周围的美丽景色。宽大的橡木地板搭配纹理细腻的大理石，营造出一种恒久高贵的气质。为中和室内过于素雅的色调，设计师特别采用了造型别致的灯具。在业主的要求下，别墅装上了许多节能和发电装置，如地热供暖系统等，以保证建筑的可持续发展。

Lennox Street

伦诺克斯街别墅

- Design Agency: Luigi Rosselli Architects
- Location: Sydney, Australia
- Photographer: Justin Alexander

- 设计单位：Luigi Rosselli 建筑设计机构
- 项目地点：澳大利亚，悉尼
- 摄影：贾斯汀·亚历山大

With these alterations and additions to an 1920's Arts and Crafts residence in Mosman we sought to modernize a family home by providing more open, family-oriented spaces such as an eat-in kitchen, rumpus room, family room and larger bedrooms, whilst maintaining the character of the original house and grand garden.

The interiors designed by Louise Bell of Interni, the landscape by Will Dangar and the architecture implemented by Candice Christensen and James Horler have all contributed to magnifying the timeless character of the house.

这是一栋位于莫斯曼的老宅。通过一系列的改建加建，设计师将其打造成了一个具有现代化特色的家庭住宅。改建后的别墅保留了房子的原始特点，保留了一个大花园，同时增设了更为开放、注重家庭氛围且具有不同功能的房间，如餐厨一体的厨房、游戏室、家庭活动室和宽敞的卧房。

本案是设计师团队合作的优秀作品，他们的巧思将这栋老宅变成了一栋现代化的别墅。静静屹立在那里，仿佛从未受时间的影响。

TOKYO
ROPPONGI
SENTOSA
HARBOUR St
HONG KONG
BOWEN Rd
NEW YORK CITY
CENTRAL PARK
COLUMBUS CIRCLE
WEST VILLAGE
SINGAPORE
BALMORAL BEACH
CLIFTON GARDENS
LENNOX St

Elm Court

榆木庭院

- Design Agency: AR Design Studio
- Designer: Mike Ford
- Location: London，England
- Photographer: Martin Gardner

- 设计单位：AR 设计工作室
- 设计师：马克・福特
- 项目地点：英国，伦敦
- 摄影：马丁・加德纳

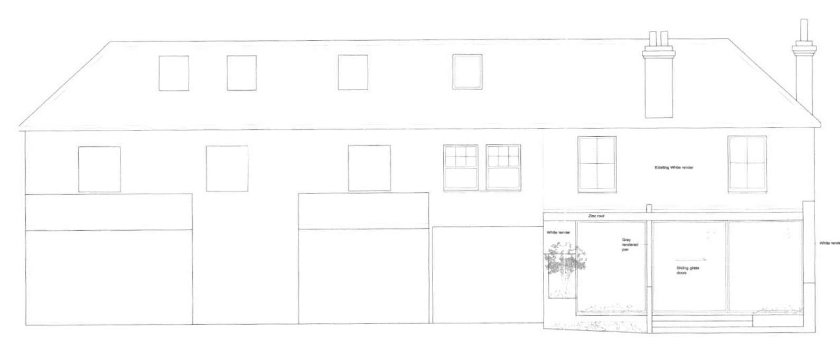

South Elevation 南面立面图

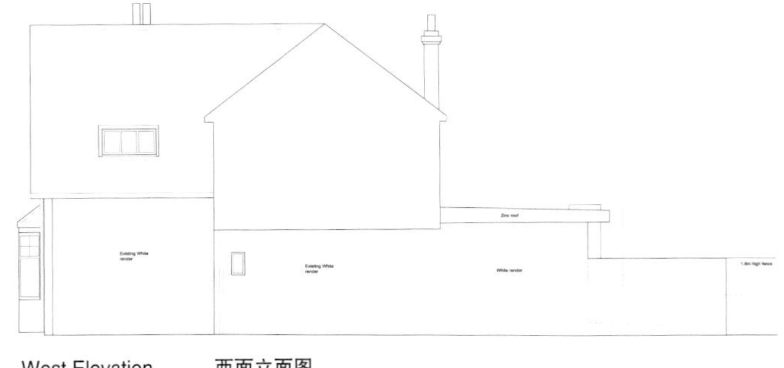

West Elevation 西面立面图

Hidden from view, behind the ordinary looking façade of a modestly sized semi-detached house in North London, sits an elegant piece of modern contemporary architecture. The recently retired owners sought to adapt the existing small and cramped property into their perfect home. The open-plan extension was to comprise of a new kitchen, new dining/living space, utility room, cinema room, WC and office area; with the aim to bring views of the garden and more light into the property.

The concept is firmly rooted in uniting the house with the garden and positively affecting the interior feel of the original property, whilst making minimal physical changes. The solution was to have a grey wall running through the entire home; from the front door through the original house into the new extension and out onto the garden to unite the original front with the new extension. This physical wall acts as an axis of movement, guiding you through the house, from old to new. It also draws the eye through the space and out to the garden providing a physical inside/outside link. As the wall reaches into the garden it projects through the large glazed doors, which can be opened completely. These allow sunlight to flood through the open-plan space and filter down gradually into the original property, creating beautiful shards of light. A flush threshold allows the tiled floor finish of the interior to continue out onto the patio, creating a seamless link between inside and outside. We follow the wall over the patio as it steps down onto the earth where it finishes, revealing us to the garden and nature.

The entire extension is highly insulated to improve the sustainability credentials of the house. In addition, an efficient fireplace is located centrally to the space for localised heating, with under floor heating running throughout the ground floor. This provides a comfortable and efficient environment to live in.

The finished property retains its humble street appearance, yet is completely transformed from its previous gloomy atmosphere with the juxtaposition of the modern extension at its rear creating a light, airy and open living environment. It is now a perfectly working home bursting with traditional values, contemporary style and innovative design, all within a very modest sized property.

本案是位于英国伦敦东部的一座半独立式洋房，在不起眼的外观之下，这座现代风格的建筑静静释放着它的优雅气息。刚退休的业主希望将他们名下这座小而窄的房产改造成一个完美宜居的家。

扩建部分包括一个新厨房、新餐厅、杂物间、放映室、盥洗室和办公区，开放式的设计旨在引入更多的自然光线和庭院风景。这个理念根植于设计之中。设计师欲将房屋和院子连通起来，从而使室内外得以沟通，并且避免过多结构上的改动。解决办法就是在屋内建起一座灰墙，从前门起，穿过屋子，一直到扩建区，最后延伸至庭院，以连通房子原始区域和扩建部分。这扇墙很好地充当了一个运动轴线，领着来人进入房中，从老宅穿越到新区；同时它还提供了一个视线向导，牵引人们的视线从室内转向庭院。玻璃门的设计，使室内变成了一个完全开放的空间，每当阳光洒入，光线逐渐晕开，一幅非常美丽的光影画面便呈现在你眼前。与地面齐平的门槛，使室内的地板缓缓延伸至天井，自然的过渡让内外空间仿如无缝联合。置身其间，沿着灰墙经过天井，再踏上院中的土地，便能与自然来个亲密接触。考虑到房屋的可持续发展，整个扩建的部分都采用了隔热材料。此外，设在中间用于局部供暖的壁炉，搭配底楼的暖气，为住户提供了一个舒适便捷的居住空间。

本案保留了街边小楼简陋的原始外观，但通过设计师的巧思，将原本晦暗的室内空间打造成了一个明亮通风的开放式的居住空间。虽然房子不大，但如今已经成为一个集传统特色、现代特点和创新设计于一体的家。

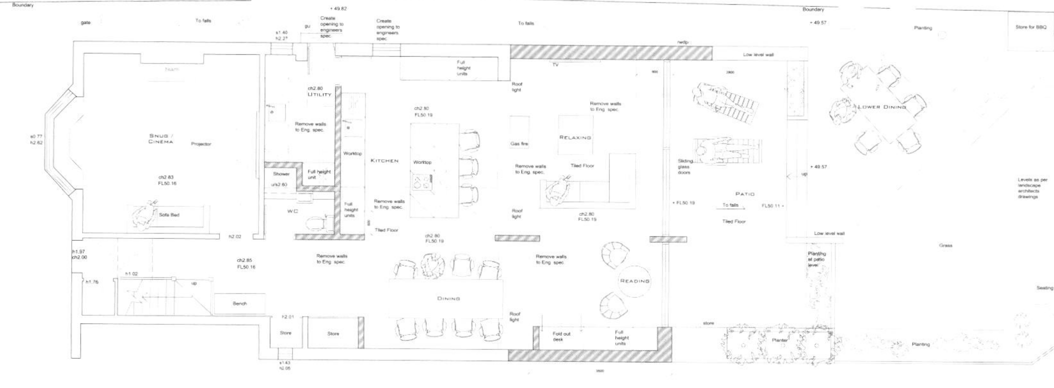

Plan 平面图

Mop House

Mop 别墅

- Design Agency: AGi Architects
- Location: Kuwait
- Area: 1300m^2
- Photographer: Nelson Garrido

- 设计单位：AGi 建筑设计机构
- 项目地点：科威特
- 项目面积：1300m^2
- 摄影：尼尔森・嘉里多

Mop House is the latest single-family housing designed by AGi architects. Originally planned to house one family with two small children, the project responds to both current and future clients' needs; on which the client plans on dividing the residence into two individual units in the future. The client wishes to live in a functional house that responds to all their daily needs. In addition, they require a public area open to the garden that leaves a distinguished impression on visitors.

Located in a residential area in Kuwait City, the site consists of a rectangular plot, which can be accessed from either side of the surrounding streets to allow for both a private and a public entrance. After moving along a curved wall that guides the visitor from the exterior of the plot into the center, one reaches the main entrance into the house. Upon entering through the main door frame, the space opens up to reveal the swimming pool and the public living areas of the house.

The garden next to the house is the vertical axis and generator of the movement of the volumes. The form of the residence is reminiscent of the movement patterns of a mop, from which flexible volumes are organized diagonally around a central axis. This axis twists upwards to generate spaces that that channel views into different directions: including the front side of the house, the side gardens, and angles of the back street.

The integration of the curve in the project responds to the movement of volumes and circulation. The circulation surrounding the patio on the first floor contrives of a succession of living spaces, which not only communicate to one another, but also relate visually the interior section of the patio to the exterior. The first floor overhangs to shade the rooms on the ground floor, and the patio is designed to define a break in between the multiple volumes of the house, which subtly reveals a side garden.

The Stairs are cladded in bamboo, marking a special space in the house. The stairs ascend leaving a hollow space in the center, from which several suspended lights are placed at different heights conveying a particular originality to the space.

ENTRANCE FLOOR
level ±0.00

The first floor plan　　一层平面图

Second floor plan　　二层平面图

Mop 别墅是 AGi 建筑设计机构设计的一栋别墅。业主育有两个子女，要求将住宅划分为两部分独立单元，以满足未来孩子对住宅的需求。同时业主希望将房子打造成一个多功能住宅，要求将房子的公共空间向花园开放，并拥有一个能给来访者留下深刻印象的外观。

本案位于科威特城的一个住宅区，地形呈矩形，可由周围小道两侧进出。沿着曲面墙进入中心区，便看到房子的主要入口。从入口进去，室外的泳池和主要生活区便呈现在你眼前。空间以屋旁的花园为纵轴，围绕中轴按斜对的方式灵活分布。沿曲折轴线分布的空间使站在通道中的人可以从不同的角度看到室内外的景象，如前边看房子，侧边看花园，后边看街道。该案利用了曲线空间的组合来优化住宅的空间流动和循环。一楼露台周围的空间设计使住宅空间得以相互交流，并在视觉上使内外空间相互连通。一楼的楼梯引人向上进入二楼，楼梯被竹板覆盖，形成了室内一个特殊的空间，吊灯错落有致，为空间增添了几许创意。

ecret House

秘密之屋

- Design Agency: AGi Architects
- Location: Shuwaikh, Kuwait
- Area: 2,600m^2
- Photographer: Nelson Garrido

- 设计单位：AGi 建筑设计公司
- 项目地点：科威特，舒伟赫
- 项目面积：2600m^2
- 摄影：尼尔森・加里多

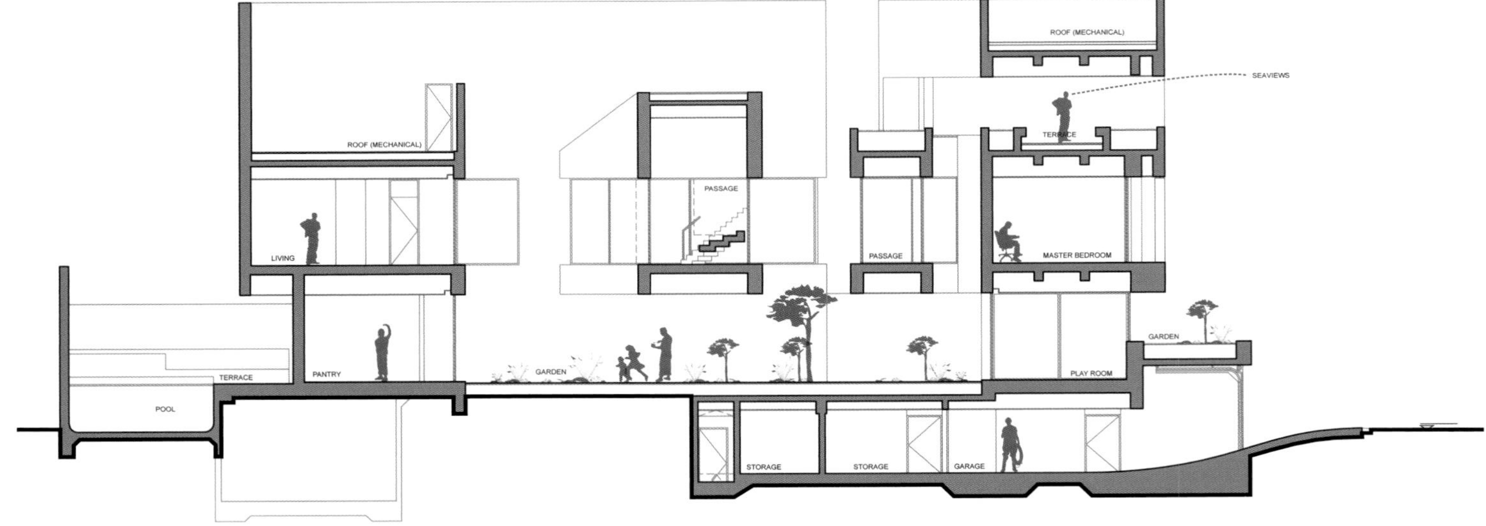

Elevation　　立面图

This house develops in different levels where different degrees of intimacy are achieved.

On the ground floor there is a garden, pool and all areas where most public activities take place. Upon entering the house we get into the main garden which is on an upper bound to the streets to ensure privacy. Once inside there are two areas; guest meeting area, as an independent pavilion, and family meeting room which connects to the pool and central garden. This area works as an open space where dining room and the open kitchen dialogue and reinforces the concept of family life.

On the first floor it has been developed more intimate spaces / private home. Here are the three rooms, located according to its size and privacy, alternates with areas of daily family living. The master bedroom has a superb location enjoying views from the main facade. The children's rooms are raised above the garden, where each unit is shown on the other respecting certain independence to preserve the family unity. As a meeting space between different rooms, there is a lounge, which extends outward to a private and secluded backyard, protected from views from the central garden.

From this level, and by a staircase facing the courtyard, you reach the second floor, where the most exclusive areas of the house are placed. Gym / massage, lounge, and large covered terrace that opens onto the main facade. The panoramic terrace is located on the roof and allows a closer relationship with the city and the sea, where enjoying a splendid view of the skyline from a private and quiet.

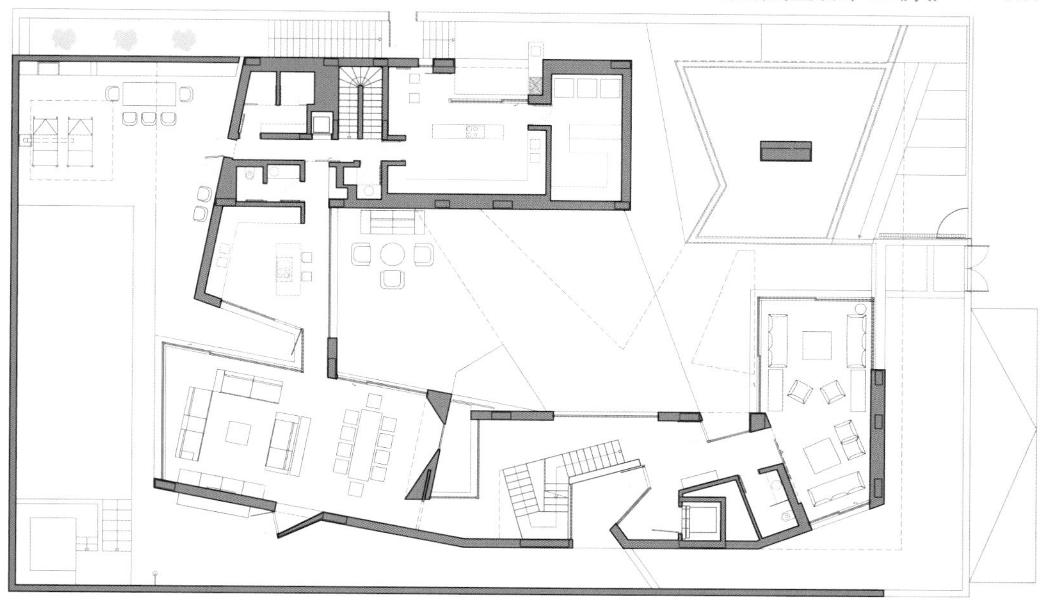

The first floor plan　一层平面图

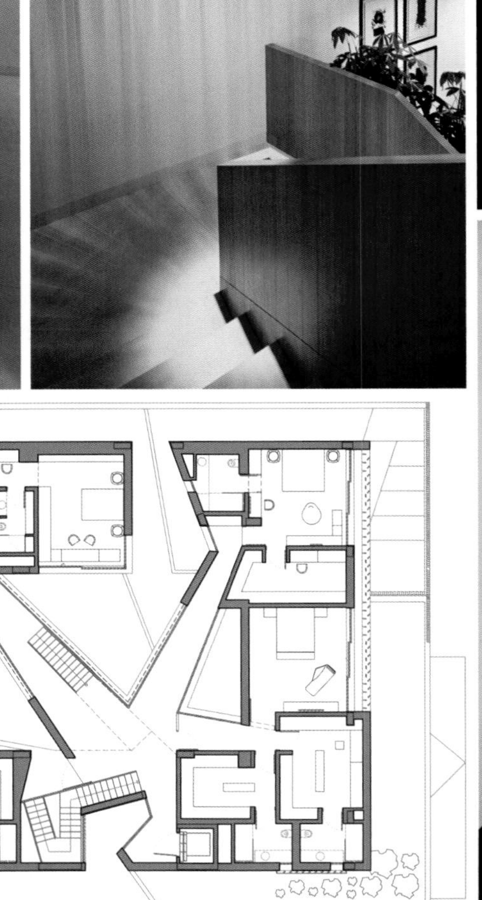

Second floor plan　　二层平面图

本案针对不同的楼层有不同的设计，且赋予各空间不同程度的私密性。

一楼户外有花园、泳池和公共活动场地。进门首先进入一楼的花园，分为两个区域，一个是类似独立亭台的会客区，一个是家庭会议室，与泳池和中心花园相连。这个开放的空间便于餐厅和开放式厨房的沟通，强化了家庭生活的概念。

二楼的设计加强了空间的私密性。三个房间分别根据空间大小和私密程度来划分，并根据家庭日常生活习惯来分配。主卧位于房屋正面，拥有极好的视野。孩子的卧室则安排在靠近花园的方位，每个房间都具有一定的独立性。房间之间的区域设置了一个可供休息闲谈的小隔间，将空间向外延伸至一个隐蔽的后院，避免站在中心花园的人直接看到房间内部的情形。

从二楼沿着面向庭院的楼梯往上走，便到达了私密性最强的三楼。在这里有健身房、按摩室、休息室和一个宽阔的平台。这个位于顶楼的平台拥有良好的视野，可以透过天窗在这静静欣赏城市风景和海景。

The Medic's House

Medics'住宅

- Design Agency: AR Design Studio
- Designer: Andy Ramus, Laurent Metrich

- 设计单位：AR 设计工作室
- 设计师：安迪・雷默斯，劳伦・米特里奇

At ground floor level the extension contains a utility, WC, kitchen, dining room and lounge area, fitted with 3 large eco-friendly sliding glass panels creating an uninterrupted view of the garden. The flush threshold and continuous floor surface enhance this connection with the garden by allowing the internal space to flow seamlessly out into it on warmer days.

The walls are constructed from super insulated block and oversized insulated cavities ensuring a very thermally efficient envelope. Large opaque glass panels to the sides allow etch light to enter deep into the plan of the space. The structure is hidden in strategically placed fins that suggest living zones within the open-plan space.

Upstairs, the western red cedar clad addition consists of a generous master suite with a separate dressing area and one other additional bedroom. This upper box is also fabricated in timber, allowing for a light weight structure that reduces the need for unsightly columns beneath. The construction contains over 250mm of insulation which AR felt was important at the upper level. This approach to construction was also carried through into the over insulated single-ply roof. The fenestration was resolved as a series of verticals that celebrate the depth of the walls with a combination of recessed and flush frameless windows.

Elevation　　立面图

设计师对本案一楼进行了扩建，规划出杂物间、洗手间、厨房、餐厅和休闲区，还安装了3面巨大的可滑动环保玻璃板，使人即使身处室内，也能欣赏院中美景。与地面平齐的门槛和由室内向外延伸的地板，使室内空间和户外庭院自然相接，在温暖的天气里，开门就能让室内外景致融为一体。墙面采用了隔热性能良好的砖石材料，以加强房子保温隔热的作用。墙上装的巨大玻璃面板并不反光，因此光线得以进入室内更深的空间。

二楼空间的墙壁贴了杉木板，外墙则采用了轻便的木料。这里有一个宽敞的配了换衣间的主套间和卧室。同一楼一样，二楼空间和屋顶也采用了环保隔热材料。设计师还在正面开了4扇无框落地窗，嵌在墙壁上，与地面垂直，透明玻璃的材质让空间显得更有深度。

The first floor plan　　一层平面图

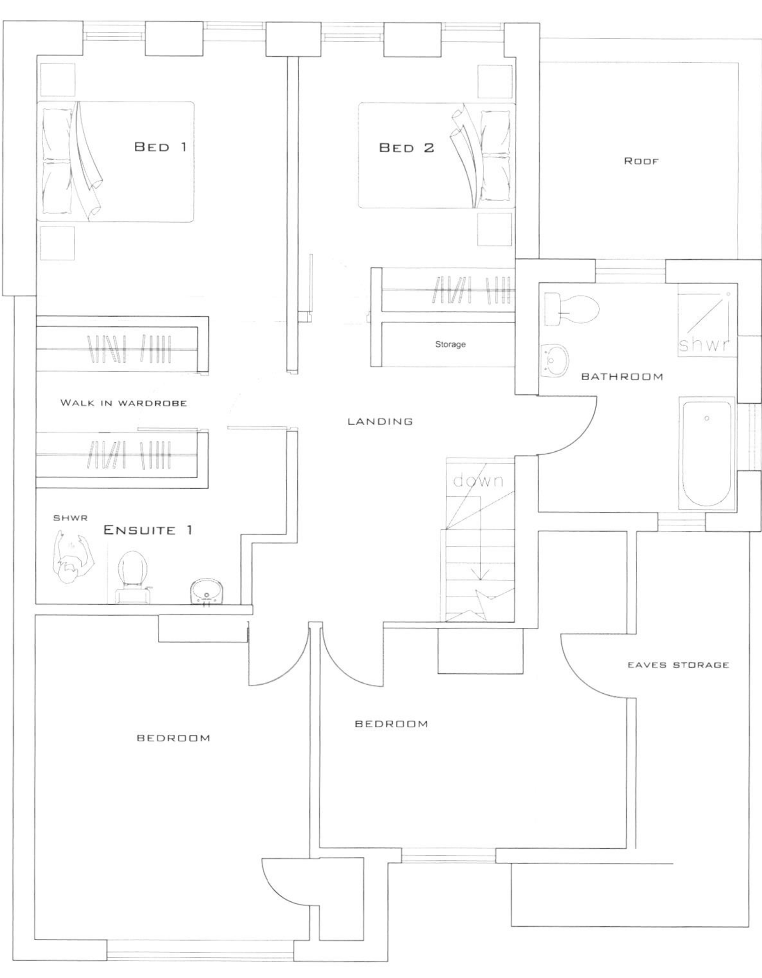

Second floor plan　二层平面图

House of California Style

加州馨居

- Design Agency: Pinki Interior Design Liu& Associates IARI Interior Design
- Designer: Danfu Liu
- Location: Xinjiang
- Area: 290m^2

- 设计单位：PINKI（品伊国际创意）& PINKI DESIGN 美国 IARI 刘卫军设计事务所
- 设计师：刘卫军
- 项目地点：新疆
- 项目面积：290m^2

This case is designed as California style, whose design idea takes into consideration the graceful scene on beaches, livable climate conditions and simple but mixed architectural and interior design characteristics of California. On stepping into the living room, you can see all the interpretation of Californian elements which can draw your eyes to the south California exotic charms.

In the living room: candlestick chandeliers on the ceiling are like tree branches; green curtains dance to the breeze whilst white sofa combines with dark coffee leather and wood to create an elegant atmosphere.

Artistic sense fills the entire space: there is a cello and an antique phonograph in the leisure book bar on the second floor. Leisure book bar is close to the balcony themed in paintings, providing a great atmosphere for art creating. There is a book shelf in every room and books either under warm lights in dining room or lying in leisure book bar silently as well as hardcover books in the main bedroom all show the reflection of modern people about the life.

本案为加州风格，设计师将加州海滩的旖旎景致、宜居的气候条件、质朴的建筑及室内造型等特质都纳入设计理念的思考。从踏入居室的大门起，随处可见加州风格元素的细腻表达，南加州的异国风情瞬间映入人们的视野。

从餐厅往里看，可望见客厅的景致：顶棚上悬挂有烛台式枝形吊灯，微风拂过，绿色窗帘随之曼妙起舞，白色的布艺沙发反衬出深咖皮革及木质的稳重。

整个空间充满艺术氛围。客厅竖立着一把大提琴，而在二楼的休闲书吧，则有老式留声机的身影。休闲书吧与绘画主题的露台相邻，为艺术创作营造了良好的氛围。每个居室中都有立柜书架，无论是客厅中暖暖射灯下的书籍，还是静静置于书吧及主卧的精装书，都折射出现代人对生活的思考。

The first floor plan　一层平面图

Second floor plan　　二层平面图

Third floor plan 三层平面图

The basement of plan 地下室平面图

Earl & Modern Spirits

伯爵·摩登情怀

- Design Agency: Shenzhen Haida Decoration · Shenzhen Wenjian Villa Design Studio
- Location: Shenzhen, Guangdong
- Area: 520m^2

- 设计单位：深圳海大装饰·深圳文健别墅空间设计
- 项目地点：广东深圳
- 项目面积：520m^2

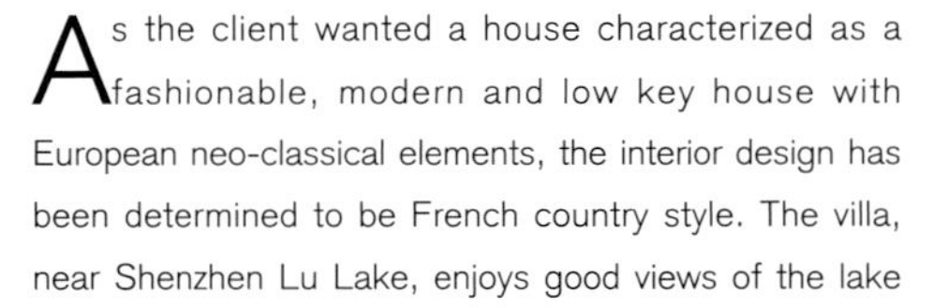

As the client wanted a house characterized as a fashionable, modern and low key house with European neo-classical elements, the interior design has been determined to be French country style. The villa, near Shenzhen Lu Lake, enjoys good views of the lake and mountains around.

In order to fulfill the business and high-end residential requirements at the same time, the designers have considerably adjusted the interior space to create a new spatial order. The interior design integrates the modern and gentle feelings with the aesthetic of naturalism, displaying elegance and solemn of neo-classicism while fulfilling the modern people's requirements for nature, regional culture and freedom.

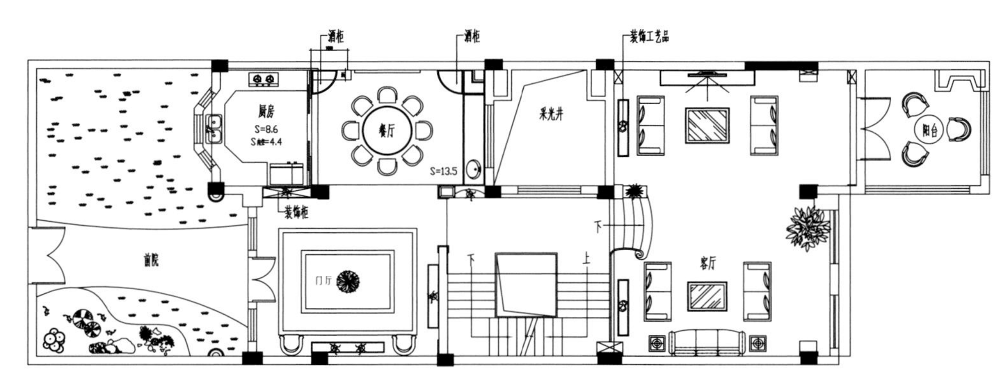

The first floor plan　　一层平面图

本案业主希望他们的家是一个时尚、现代、独特、低调，又有一定欧式新古典元素的家。分析了一下此项目，其风格是法式乡村风格，项目功能定位为高端私宅，坐落于深圳鹭湖边上，视野开阔，可以轻松享受周边的山景。设计师对建筑内部空间关系作了很大调整，重新塑造了新的空间秩序，满足本案的商务需求定位和现代人居住宅的高端需求，使整个空间综合现代感、柔和感、自然主义的美学精神，把新古典的典雅、大气、庄重之感表现出来，满足当代人对于自然、地域文化、自由精神的内心需求。

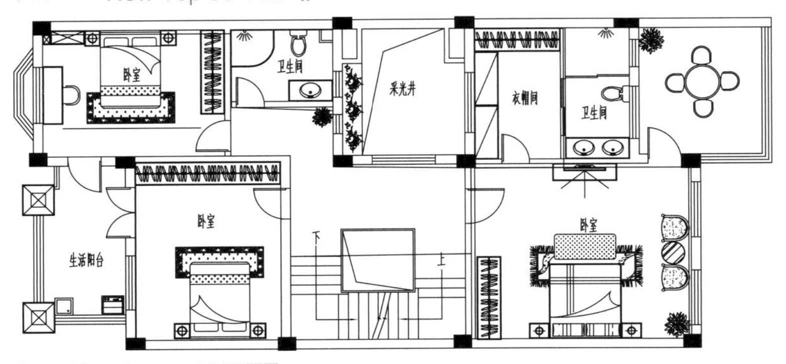

Second floor plan　二层平面图

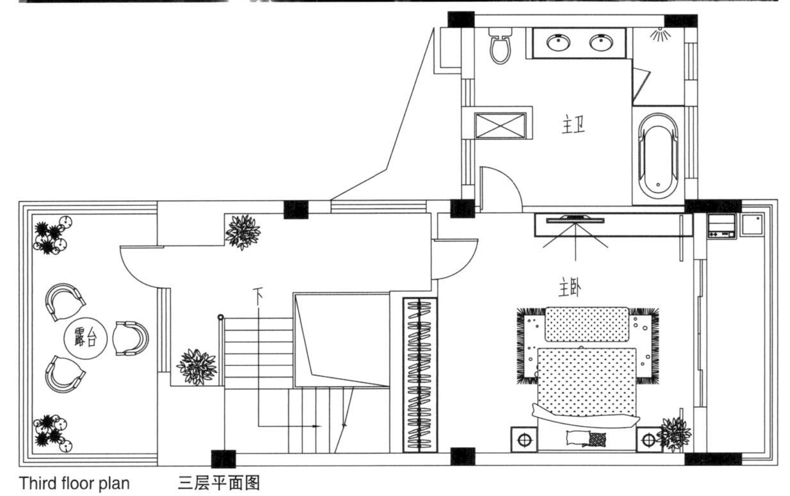

Third floor plan　三层平面图

The basement of plan　　地下室平面图

The first floor plan　　一层平面图

A Crafted Jade – Elegant Chinese Style

璞玉得琢 · 唯美中式

- Design Agency: Shenzhen Haida Decoration · Shenzhen Wenjian Villa Design Studio
- Location: Shenzhen, Guangdong
- Area: 550m^2

- 设计单位：深圳海大装饰 · 深圳文健别墅空间设计
- 项目地点：广东深圳
- 项目面积：550m^2

A raw jade needs to be carved to show its real charm. So is a good house. As the client has a preference for traditional Chinese style, the designers decided to make a concise Chinese style by using modern materials like a silver mirror, carved tawny glass, and so on, to avoid the sense of oppression of traditional Chinese style. The wall with hair surface echoes to Chinese furniture, creating an atmosphere of Chinese culture. Elements of traditional culture are used in the interior design, such as the porcelain decorated with colored drawing flowers, wood carving, silk painting, etc. Especially the three jade sculptures highlight the whole design. In traditional Confucian culture, it's meaningful to give up something in a right time, which can be manifested in the design. The client has spent considerable energy and money on the expensive mahogany furniture, and the interior design emphasizes in highlighting the atmosphere decorated by the furniture, instead of making it an eye catcher.

璞玉得琢——好的房子也需要精雕细琢。本案业主非常喜好中国传统的装修风格，而传统中式往往会使人产生繁琐、压抑的感觉。设计师以简洁中式的设计风格定位，配合现代的用材，如银镜，茶镜雕刻等手法，营造出空间的灵动。采用毛面材质做的主题背景与中式家具细腻的符号互相衬托，达到肌理上的多变与相容。

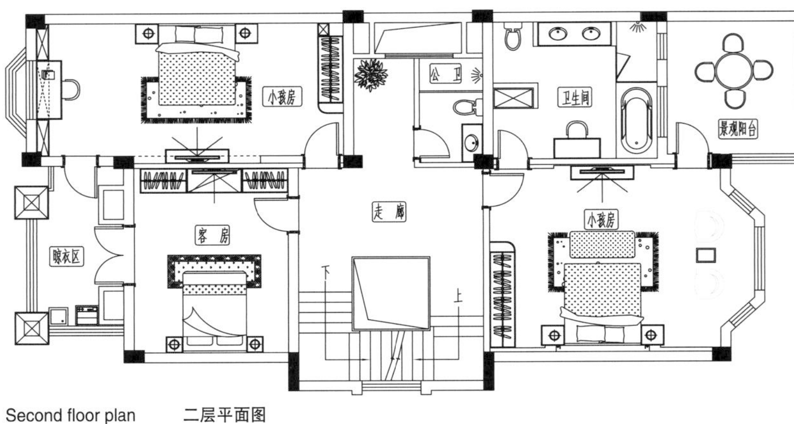

Second floor plan 二层平面图

中式风格处理不好，容易使人乏味，产生审美疲劳。这就需要在设计中加入一些传统文化的元素，因为只有文脉，才不会被历史摒弃，设计才会永恒。设计师通过运用“彩绘花瓷”、“木雕花”、“绢画”等要素加以烘托，特别是运用三件精美的“玉雕”，起到了画龙点睛的作用。

中国儒家思想中，“懂得舍弃”是重要的思想之一。设计师在本案定位中加以很好的运用。业主在红木家具投入了很大的精力和资金，设计师在室内设计中，以烘托家具的整体感觉为主，在硬装上并没有“喧宾夺主”。

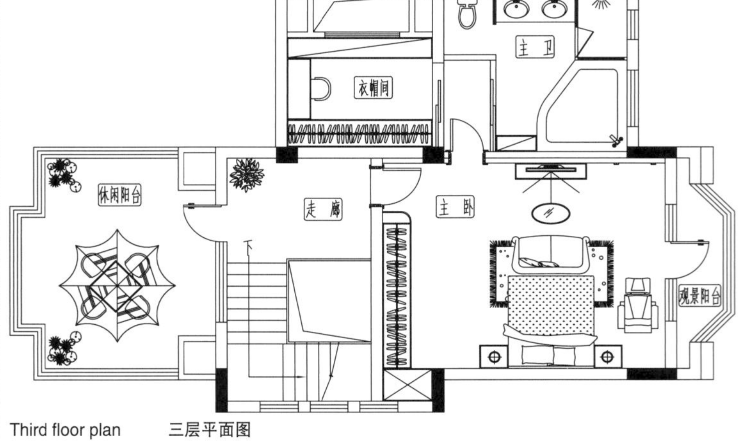

Third floor plan 三层平面图

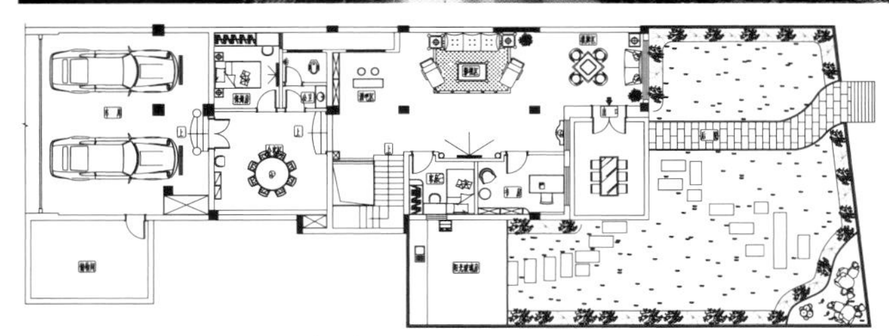

The basement of plan 地下室平面图

Lang Yi Feng House

朗逸峰

- Design Agency: Simon Chong Design Consultants Ltd.
- Designer: Simon Chong
- Location: Hong Kong
- Area: 530m^2
- Photographer: Ye Jingxing

- 设计单位：SCD(香港)郑树芬设计事务所
- 设计师：郑树芬
- 项目地点：香港
- 项目面积：530m^2
- 摄影：叶景星

The Cairn Hill house is located on the hilltop, overlooking the Gunyulgup valley, the great dam and the lake, which is said to be the best relaxing place for many famous stars and celebrities.

The penthouse enjoys a great geographical location. For it is near the hilltop in the west, the hillside and ocean view, as well as the view of Hong Kong Island, provide a great visual feast. The design takes advantage of the high raised roof whilst the colored carved lines of European classic style are used to decorate the wall. The palette is pure white to make the space more spacious and brighter visually, whilst enhancing the whole space. Some parts of the wall are decorated with elegant European wallpaper, to avoid tedium and add more freshness to the space.

On the rooftop there is a vertical garden, along with various furniture which are dividing the area into several parts, to show different functions, where the owner can invite friends to have a party.

朗月峰宅基地位于山顶，占地 10 英亩，可俯瞰 Gunyulgup 谷，大坝、山谷和湖泊一览无遗。据悉，此处是香港诸多明星和社会名流度假休闲的好去处。

此套顶层复式别墅地理位置非常优越，由于靠近西部山顶，其室外景观一览无遗，同时拥有山景和海景，加上部分港岛，形成壮丽景色。

该顶层复式别墅利用了楼顶高挑的优势，将欧式古典花线围身作为主要装饰面板，特别是整个空间以纯白色为基调，在视觉上宽敞而明亮，彰显高雅大方的气质。部分位置利用典雅丰富的欧式墙纸作为点缀，让每一个空间都会有变化，给人焕然一新的感觉。

顶层复式还有另外一个优点便是天台，宽敞的天台安排了 18m 宽的垂直花园作为重点来装饰，不同的户外家具也将整个天台进行区分，展现出不同的功能氛围，业主可尽情邀请朋友 Party，充分享受生活！

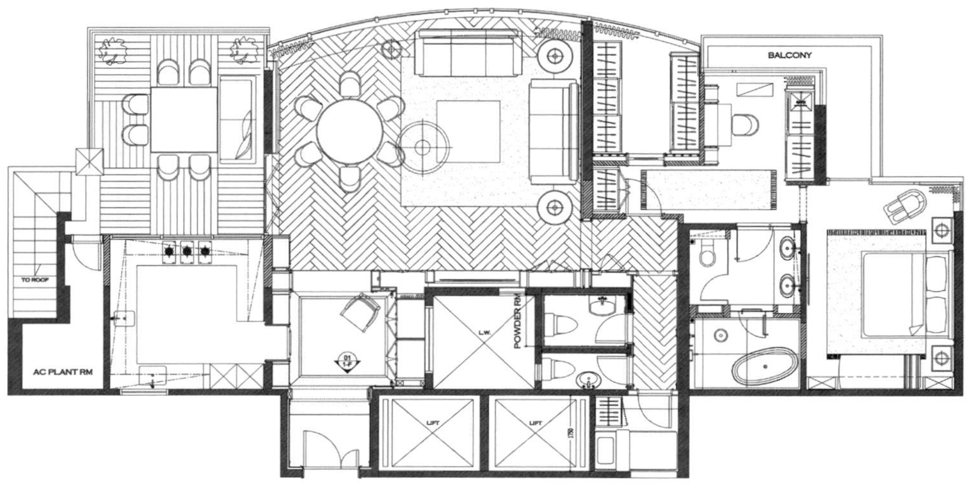

Plan 平面图

Guangzhou South Lake Villa

广州南湖山庄

- Design Agency: Danny Cheng Interiors Ltd
- Designer: Danny Cheng
- Location: Guangzhou, Guangdong
- Area: 353m^2

- 设计单位：Danny Cheng Interiors Ltd
- 设计师：Danny Cheng
- 项目地点：广东广州
- 项目面积：353m^2

The four-storey house is located near the South Lake Scenic Area. To blend into the natural environment, the house has been designed as a casual holiday house. The designer takes advantage of its height and provides a fashionable and unique design to both the interior and the exterior. Looking from the distance, it seems to be far away from the bustling city.

The batten revolving door and screen, the well-arranged balcony and a cantilever pool, and many geometric design details add unique personality and connotations to the mid-hill villa. Thanks to the ground floor's height, the whole building has a special architectural charm. Complying with the natural minimalism, the designer maximizes the interior design's elements, colors, illuminations and original materials. However, it requires good colors and material textures, which restrains the brief-doctrine space design.

At the entrance, the ground and walls decorated with black and grey marble, echoing with log wooden floor, add natural feels to the design. When sunlight comes in through the window, it will be reflected from the marble walls to the wooden floor, creating beautiful flowing lines, whilst nourishing the green plants, to create a natural atmosphere and help the owner to switch its roles freely.

这是一座 4 层高的别墅，位于著名的南湖风景度假区旁，设计师把本案打造成了悠闲写意的度假屋以融合周边翠绿茂密的自然环境。在高度上拥有得天独厚的优势，再加上 Danny Cheng 赋予其时尚特别的内外造型，两种元素在设计中交织融化。在远处观望，这座山顶大宅给人远离城市喧嚣的距离感。

设计师以简洁的线条、独特的木条子旋转门、木条子屏风、错落有致的阳台及空中悬挑泳池等几何空间独特设计，赋予半山建筑独一无二的态度和内涵。迈进大门的一霎那，重新审视这座别墅，你会发现在它高傲的外表下掩藏的是一种富有亲和力的时尚感，因楼底特高，看起来充满建筑感。

室内设计遵循自然简约主义，设计的元素、色彩、照明、原材料都被设计师简化到最少的程度，但这种设计风格对色彩、材料的质感要求很高，所以简约的空间设计非常含蓄。入口处，设计师以灰色和黑色的大理石点缀地面、墙面，配以木质原色的地板使得整个建筑设计充满了自然的味道，同时这两种不同材质的搭配也点亮了整个空间。阳光从窗户投射进来时，光线会从大理石的墙面反射到木质的地板上，形成流动的线条，滋养旁边的绿色植物，使整个空间充满自然的气息，同时也能让业主自然地进行角色的转换。

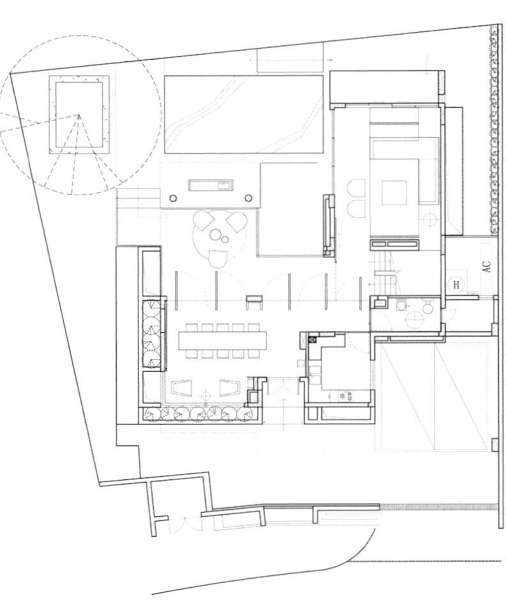

The first floor plan　一层平面图

Second floor plan　　二层平面图

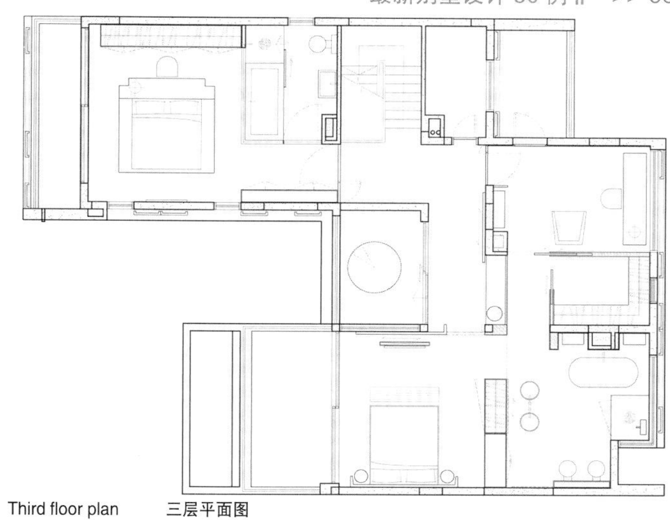

Third floor plan 三层平面图

Terrace House in Paddington

帕丁顿联排住宅

- Design Agency: Luigi Rosselli Architects
- Location: Sydney, Australia
- Photographer: Justin Alexander

- 设计单位 : Luigi Rosselli 建筑设计机构
- 项目地点：澳大利亚，悉尼
- 摄影：贾斯汀·亚历山大

A designer would find oneself dancing to a familiar tune when approached to upgrade this terrace house in Paddington, a suburb east of Sydney City.

Faced with the age old problems presented by much loved terrace housing – damp, dark and introverted – we sought to create a luminous space to give a full family a much needed dose of vitamin D. Introducing some fluid lines with a light filled stairwell at the centre and a sun drenched kitchen and living at the rear, the new configuration of old and new proves an enriching experience. Accustomed to muted tones, and a subtle palette, a much needed spring was put in our step by the bold use of colors, delphinium blues, cadmium yellows, beautiful artworks, exotic patterns and rich textures carefully selected by the interior designer in residence, Heidi Correa. The lush landscaping at the rear provides a verdant backdrop to family life. The final result knocked even us off our feet.

本案是一栋位于悉尼东部城市帕丁顿的联排住宅。

在本案的设计中，设计师也遇到了大部分联排住宅都有的老问题：潮湿晦暗不通风。因此，设计师欲将本案打造成一栋明亮通风的别墅，可供居住家庭从日光中获取重要的维生素 D，从而享有更明亮健康的生活。房屋中间的楼梯光线充足，屋后的厨房和客厅沐浴在阳光下，房子的设计处处体现着改造前后的优点相融相和的特色。原本房子的色调偏浅淡柔和，因此设计师在用色上大胆突破，采用翠雀蓝、镉黄等亮丽的颜色，并用具有异域风情的工艺品以及丰富的花色和纹路来装饰整个空间。后院绿意葱茏的植物景观，为业主提供了一个充满自然生气的生活背景。经过设计师的巧手改造，老房一改黯淡旧貌，成功换上惊艳新颜。

17 BR-House

布莱尔路 17 号别墅

- Design Agency: ONG&ONG Pte Ltd.
- Location: Singapore
- Area: 404m²
- Photographer: Aaron Pocock

- 设计单位：ONG&ONG Pte Ltd.
- 项目地点：新加坡
- 项目面积：404m²
- 摄影：亚伦・波科克

Elevation1 立面图 1

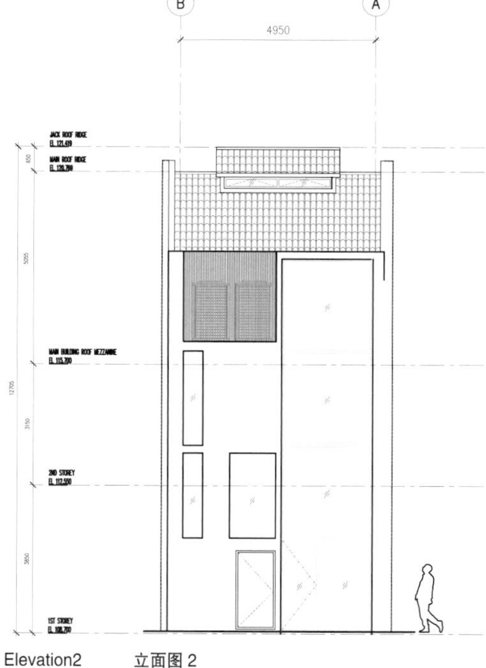

Elevation2 立面图 2

17 BR-House is located within the Blair Plain Conservation area. The owners, a French Singaporean family, were deeply involved in numerous aspects of the conservation project. What they wanted was a warm family home that would also display much of its historical character.

Prior to conservation works, the ground floor was what remained of a long office unit. Bound by two party walls, the result was a very poorly-lit mono-space that stretched all the way to the back of the site. To bring in much needed light, a courtyard was created through which sunlight and wind could enter and naturally ventilate the house. Since the roof of the second storey had been restored to its original shape, the architects endeavoured to add a wall at the end of the roof so as to mimic the front half of the house at the back to form a symmetrical envelope. A balcony was constructed for the second floor's rear bedroom, stretching right to the horizontal length of the roof. On the third storey, there is also a skylight in the master bathroom to further illuminate both this room as well as the second floor balcony. And similar to the first floor's incorporation of historical elements, timber beams were installed into sections of the roof. The greenwall also reaches up to this level for a touch of nature. The rear block comprises of the kitchen, as well as the 7-metre long swimming pool and service quarters. In terms of conservation, traditional glazed tiles line the floor whilst a replica geometrical service staircase at the back are reminiscent of the shophouse's early days. The staircase leads all the way to the top where the servant's room and pool area are located.

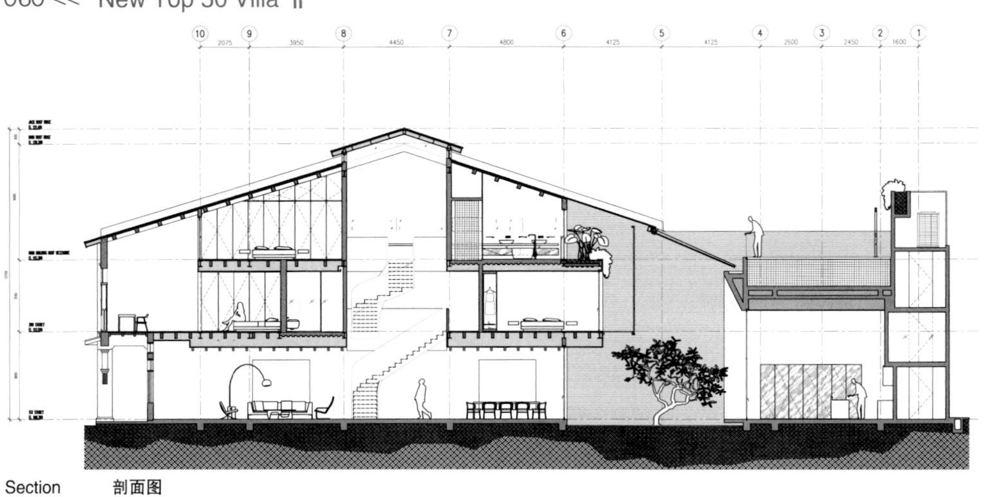

Section 剖面图

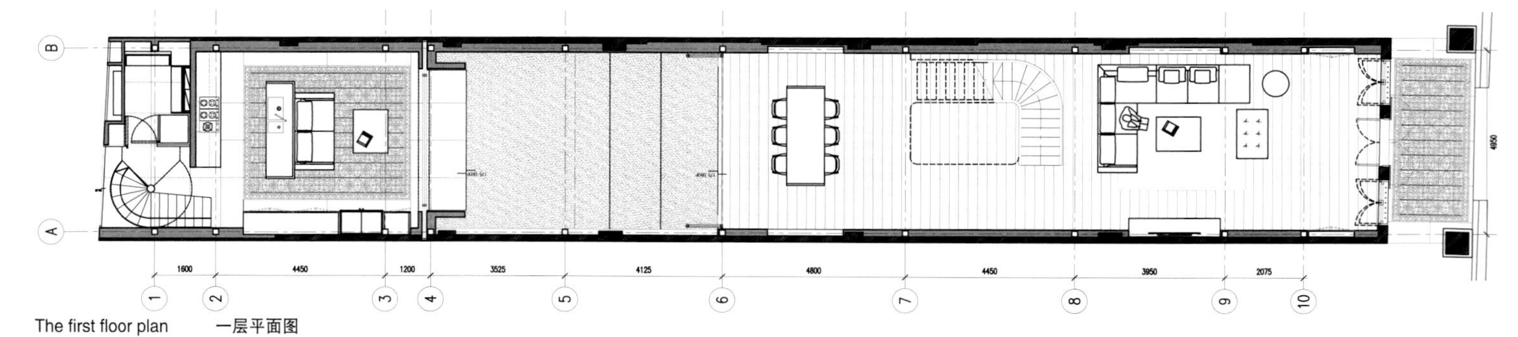

The first floor plan 一层平面图

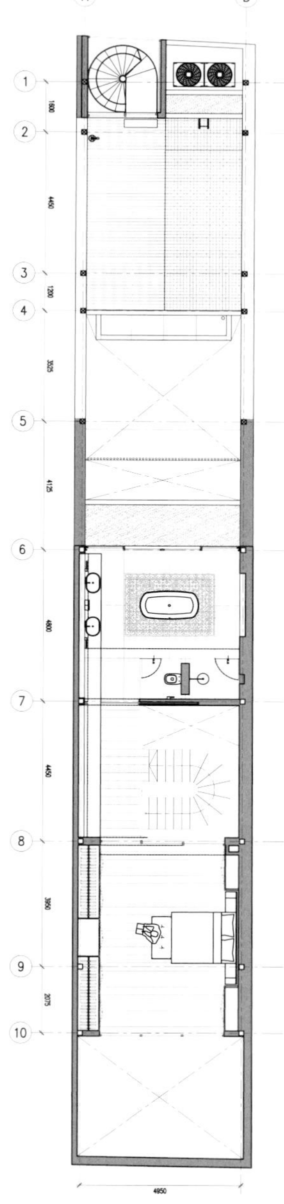

The mezzanine floor plan　夹层平面图

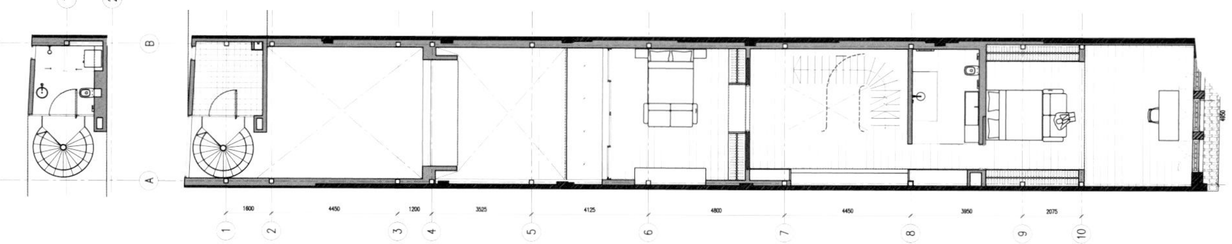

Second floor plan　二层平面图

布莱尔路 17 号屋位于布莱尔历史街道保护区。业主是一户法籍新加坡家庭，曾亲身体验过该区历史街道保护项目的方方面面。他们提出，想将房子打造成一个温暖的家庭式房屋，并且能够尽可能地展现它的历史特点。

本案的一楼曾是一个办公单元房。由两面界墙相连，室内形成了光线极差的单一空间。为了改善室内的采光和通风条件，设计师打通了后院，让阳光和自然风能够进入室内。二楼屋顶保留了原始面貌，设计师还仿照房子前面的构造，在末端加了一面墙，使之形成一个前后包覆的对称形状。阳台建在二楼后部的卧室外，长度与屋顶齐平。三楼的主卧开了一扇天窗，使得光线得以照进室内，

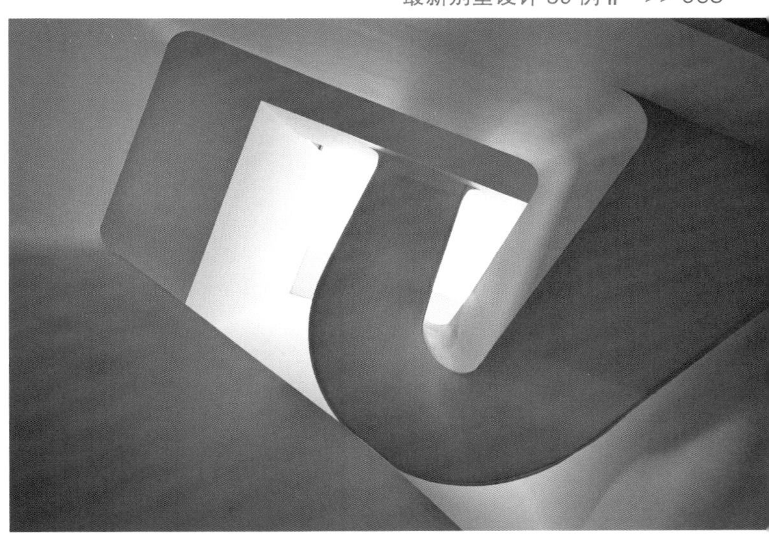

甚至延伸至二楼阳台。和一楼一样，三楼的室内设计结合了原有的历史元素，在顶棚装了一排木梁。

绿色植物墙的设计让人更加亲近自然。

房子后部的区域则由厨房和一个 7m 长的泳池及其他服务区域组成。为了保存房子的历史特色，地板铺以传统特色的光滑瓷砖，而房子后部通往佣人房间和泳池区域的楼梯，其抽象几何的形状，则是对早年曾用作商铺的房子的一种怀恋。

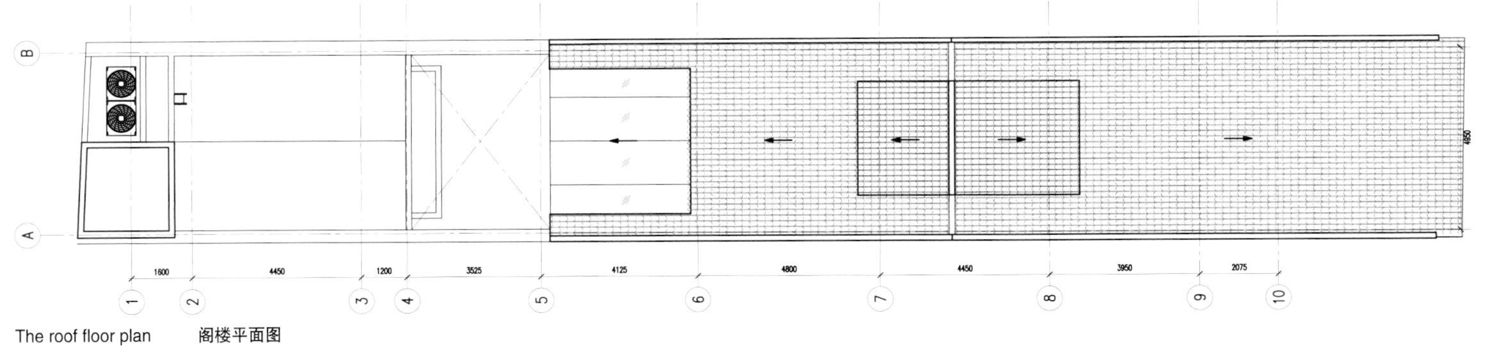

The roof floor plan　　阁楼平面图

Home 09

Home 09

- Design Agency: i29 Interior Architects
- Designer：Paul de Ruiter
- Location: Bloemendaal, Netherlands
- Area: 489m^2
- Photographer: i29 Interior Architects

- 设计单位：i29 室内设计机构
- 设计师：保罗・德・瑞特
- 项目地点：荷兰，布卢门达尔
- 项目面积：489m^2
- 摄影：i29 室内设计机构

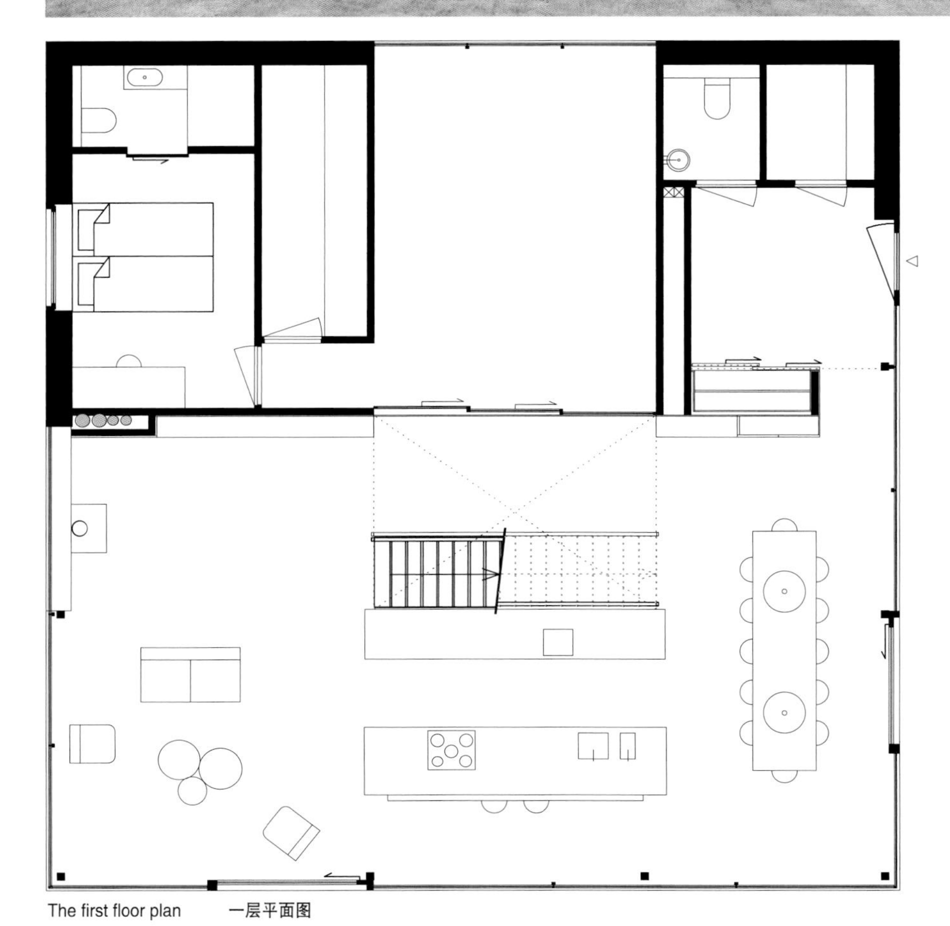

The first floor plan　　一层平面图

Close to Bloemendaal, on the edge of the Kennemer dunes, the site of Villa Bloemendaal is situated. A sustainable home that follows a minimalistic design and shows respect for man and nature alike, in a unique residential area where the existing flora and fauna are given full rein.

i29 interior architects worked on the interior of a villa which was designed by Paul de Ruiter architects. A minimal approach to the materialisation and detailing of the building is a core value of both the interior and exterior design. The large expanses of glass and the patio result in maximum daylighting and give the inhabitants the feeling that the villa and the surrounding landscape are one.

In order to bring nature inside even more, all of the interior functions in the house are made from natural materials. i29 interior architects created large surfaces of wood through the whole house to connect the different areas. Cabinets, wardrobes, walls, sliding doors, beds and even a fire place have been made in one and the same material. Pine wood panels -normally a low tech material- has been used and fabricated in a high tech and detailed way.

Home 09 是位于荷兰城市布卢门达尔附近一个植被茂盛的住宅区的一栋别墅。该别墅是一栋可持续性建筑，遵循极简主义，尊重人与自然的和谐共存。该案是由 i29 室内设计建筑机构的保罗・德・瑞特建筑师主创设计，其室内外设计的核心理念是追求极简的结构细节。

Second floor plan　二层平面图

宽大的玻璃落地窗和天井让日光得以涌入室内，实现了采光的最大化，让居住者感受到别墅与周围环境的融合。为了给室内空间带来更多的自然气息，房子的设计采用了天然材料。设计师采用大面积的木材装饰整个室内空间，连接不同的区域。橱柜、衣柜、墙面、滑动门、床，甚至是壁炉都用统一的松木板覆盖。这些本属于低科技材料的松木板通过高科技以更为细致的方式被应用于本案的设计之中。

Lake House

湖畔别墅

- Design Agency: SAOTA, ARRCC
- Location: Geneva, Switzerland
- Area: 1553m²
- Photographer: SAOTA

- 设计单位：SAOTA 建筑事务所，ARRCC 室内设计机构
- 项目地点：瑞士，日内瓦
- 项目面积：1553m²
- 摄影：SAOTA

On either side of the 20 meter wide channel sits the two portions that make this house, the main house and the annex. What link the two buildings are the cinemas, spa, auditorium and garages underneath. The main house is a combination of round edged cubes and triangular masses that form the L-shape of the living spaces. A double volume living area with a curved wall on the façade facing the lake, flows into a dining area and kitchen on the ground floor and bedrooms, a lobby and en-suite's on the top level. The top floors are accessed by a glass cylinder encased lift.

The living room is divided into two zones, the formal area and the informal arrangement centred around the feature fireplace – a black suspended flue and fire dish mounted on the floor. Continuity of these two zones was achieved by specifying the same modular sofa, the curved Arne sofa from B&B Italia, but in different configurations. Custom-sized organic-shaped patchwork Nguni rugs were designed for both of these areas.

LEGEND

1. DRIVEWAY
2. GARAGE
3. STORE
4. OFFICE
5. THEATER
6. POOL
7. STEAM ROOM
8. SAUNA
9. TERRACE
10. STAFF ROOMS
11. ENTRANCE GATE
12. POND
13. LOUNGE
14. DINING
15. KITCHEN
16. SCULLERY
17. BEDROOM
18. DOUBLE VOLUME

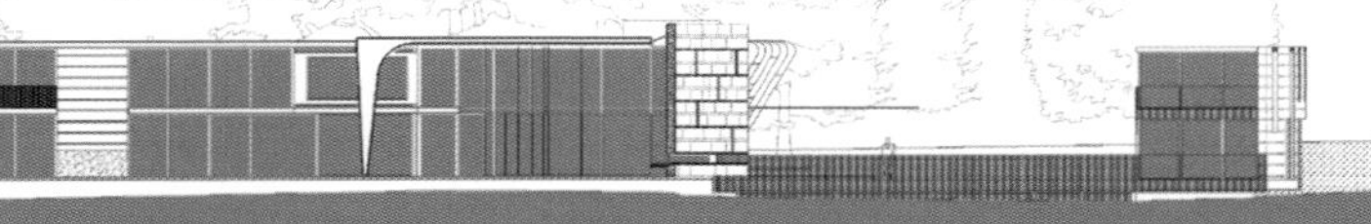

South East Elevation 东南立面图

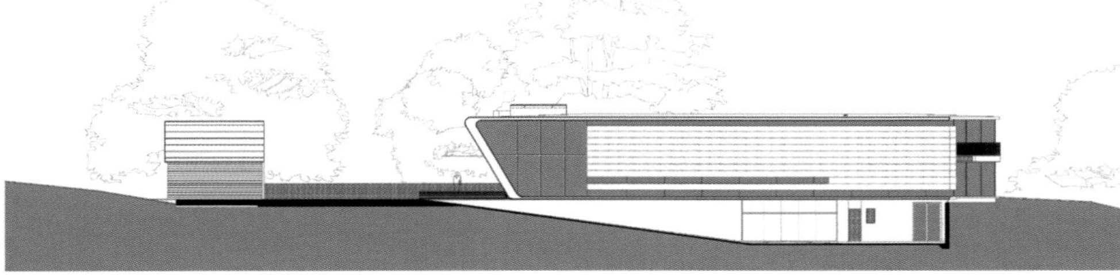

North West Elevation 西北面立面图

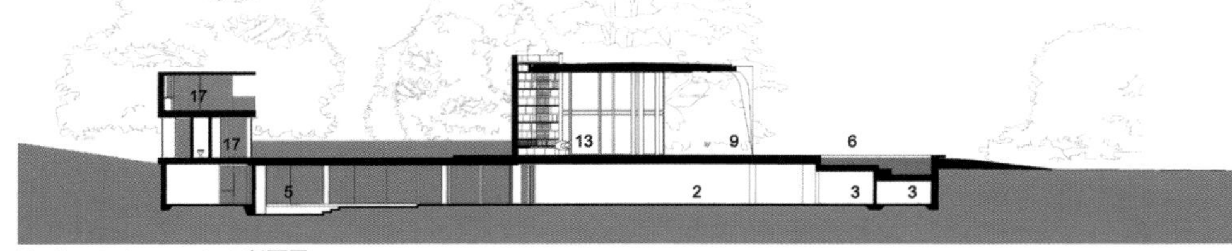

Section A–A A–A 剖面图

本案别墅分主屋和配楼两个部分。将两栋建筑连接起来的区域有电影院、SPA 馆、礼堂和地下车库。主屋呈 L 形走向，是由一些立方体和三角形空间组合而成。

两层的居住空间外有一面造型奇特的弧形墙，整个建筑面向大湖，一楼有餐厅和厨房，二楼有会客室和套间。通往顶楼可乘玻璃外墙的升降电梯。客厅划分为正式会客区和休闲活动区。在休闲活动区，中间有一个引人注目的壁炉，由一根悬吊的黑色烟道和安装在地面的火盆组成。设计师通过在两个区域以不同的方式放置相同款式的 Arne 沙发组合，来实现空间的协调统一，同时还在两组沙发前铺上了特别定制的 Nguni 拼接地毯。本案还采用了许多对应物来完成设计，比如雕塑艺术和手工工艺、天然材料和高科技应用的对比。

LEGEND

1. DRIVEWAY
2.GARAGE
3.STORE
4.OFFICE
5.THEATER
6.POOL
7.STEAM ROOM
8. SAUNA
9. TERRACE
10. STAFF ROOMS
11. ENTRANCE GATE
12. POND
13. LOUNGE
14. DINING
15. KITCHEN
16. SCULLERY
17. BEDROOM
18. DOUBLE VOLUME

The first floor plan 一层平面图

LEGEND

1. DRIVEWAY
2.GARAGE
3.STORE
4.OFFICE
5.THEATER
6.POOL
7.STEAM ROOM
8. SAUNA
9. TERRACE
10. STAFF ROOMS
11. ENTRANCE GATE
12. POND
13. LOUNGE
14. DINING
15. KITCHEN
16. SCULLERY
17. BEDROOM
18. DOUBLE VOLUME

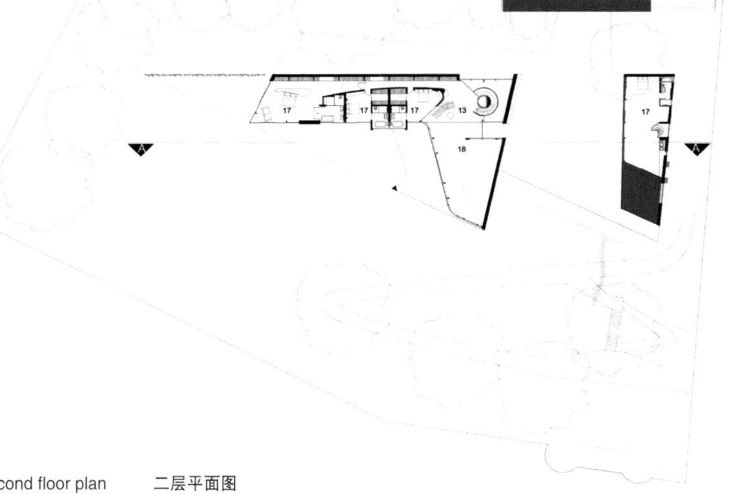

Second floor plan　　二层平面图

Sandhurst Towers

桑赫斯特公寓

- Design Agency: SAOTA,OKHA Interiors
- Location: Sandton, Johannesburg
- Area: 1000m^2
- Photographer: Dook

- 设计单位：SAOTA 建筑事务所，OKHA 室内设计机构
- 项目地点：南非，约翰内斯堡
- 项目面积：1000m^2
- 摄影：杜克

The client's brief for this three-level Sandton penthouse was to design a product that would showcase the evolution of 21st Century urban living, reflect individualism as well as evolve the technological aspects of the 'smart home', with tailored lighting and automation.

LEGEND

01. ENTRANCE
02. LIBRARY
03. DINING AREA
04. MEDIA LOUNGE
05. FAMILY LIVING AREA
06. OPEN PLAN KITCHEN
07. SCULLERY
08. STAFF QUARTERS
09. WALK IN FRIDGE
10. BATHROOM

The first floor plan　一层平面图

Court(the designer) saw this project as the establishment of a new standard in modern urban luxury living. An opportunity to illustrate how a family unit – there are four bedrooms – can live in central Sandton in complete calm, security and extreme comfort. Despite the fact that this was an 11th floor penthouse apartment in a commercial city centre, he elected to provide an unobstructed indoor / outdoor flow not normally associated with metro living.

The designers needed to maintain a fluid connectivity throughout but also give each zone its own very specific look, not to have each simply a continuation of the previous environment. This can be difficult when walls don't compartmentalise and delineate areas and to overcome it they employed stylistically different furniture and décor elements from space to space and also used color accents specific and unique to each sector.

The overall palette throughout the entire 1000m2 is black and white with additional color used as an exotic accent. Artwork, sculpture and a large collection of glassware, ceramics and other accessories play a vital role in layering the space and adding soul, depth and character.

应本案业主要求，这套位于桑顿的三层样板房应被设计成拥有定制灯具和自动化应用，且能够展现 21 世纪都市生活、个性以及智能家居技术的房子。设计师认为，本案确立了现代城市奢华生活空间的标准。

本案位于桑顿的中心，拥有 4 个卧房，为住户提供了一个舒适安逸的住宅空间。尽管房子位于商业城市中心一栋大楼的第十一层，但通过设计师的精心规划设计，房子被打造成一个室内外自然相通的住宅。本案的设计既要保证各空间的和谐一致，又要使其具有各自的特色，而不是单纯地延续之前的风格。因此在空间划分和设计上就具有一定的难度。为了解决这一问题，设计师采用了时尚各异的家具和装饰元素来装饰各空间，还着重强调了色彩的搭配以体现空间的独特性。黑白是本案的主色调，同时还采用了一些具有异国风情的色彩。工艺品、雕塑以及一系列玻璃器皿和陶瓷制品等配饰，在室内设计中起着非常关键的作用，赋予整个空间深度、个性和灵魂。

LEGEND

01. ENTRANCE
02. LIBRARY
03. DINING AREA
04. MEDIA LOUNGE
05. FAMILY LIVING AREA
06. OPEN PLAN KITCHEN
07. SCULLERY
08. STAFF QUARTERS
09. WALK IN FRIDGE
10. BATHROOM

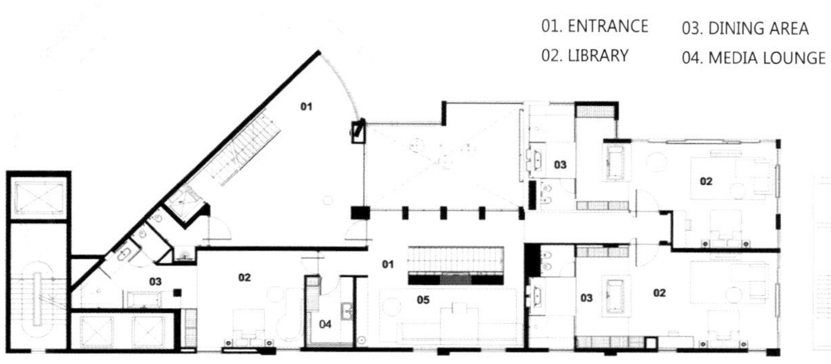

Second floor plan 二层平面图

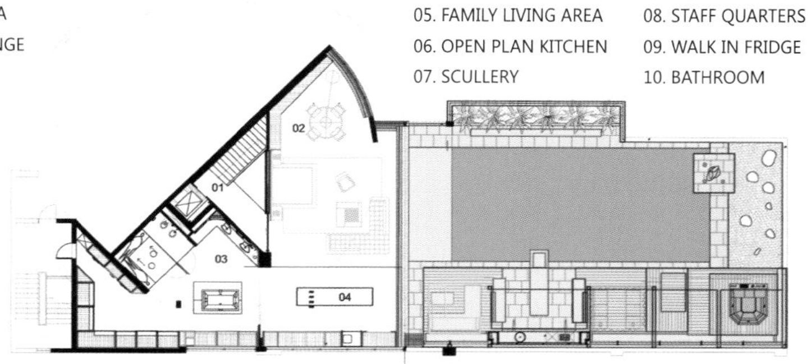

Third floor plan 三层平面图

Silver Bay

银湾别墅

- Design Agency: SAOTA, ARRCC
- Location: Shelley Point, West Coast, South Africa
- Area: 510m^2
- Photographer: SAOTA, Adam Letch & Enda Cavanagh

- 设计单位：SAOTA 建筑事务所，ARRCC 室内设计机构
- 项目地点：南非，西海岸
- 项目面积：510m^2
- 摄影：SAOTA 建筑事务所，亚当·莱奇，恩达·卡瓦纳

The site is positioned at Shelley Point on the West Coast Peninsula. Uniquely for the West Coast, the site faces East over the bay looking towards the mountains behind the small Swartland town of Aurora.

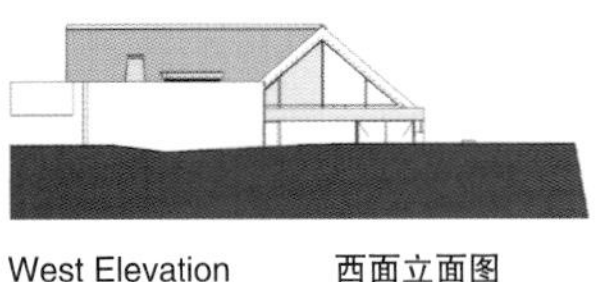

West Elevation 西面立面图

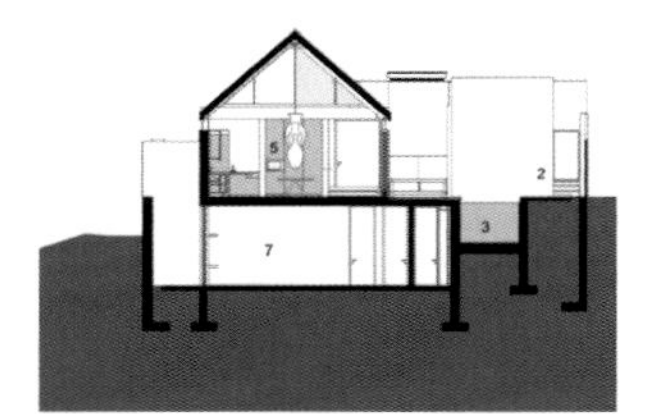

Section A-A A-A 剖面图

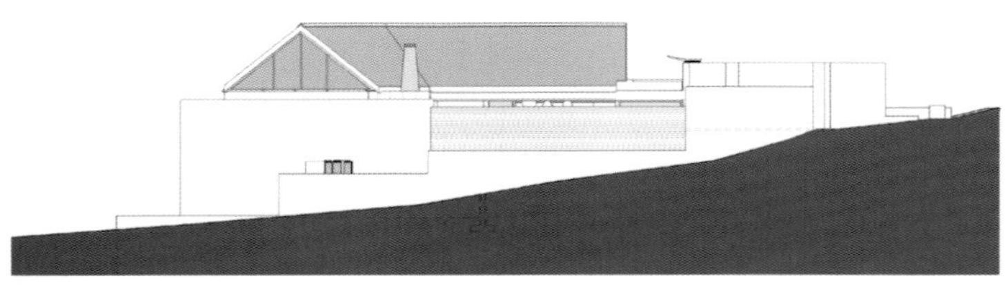

North Elevation 北面立面图

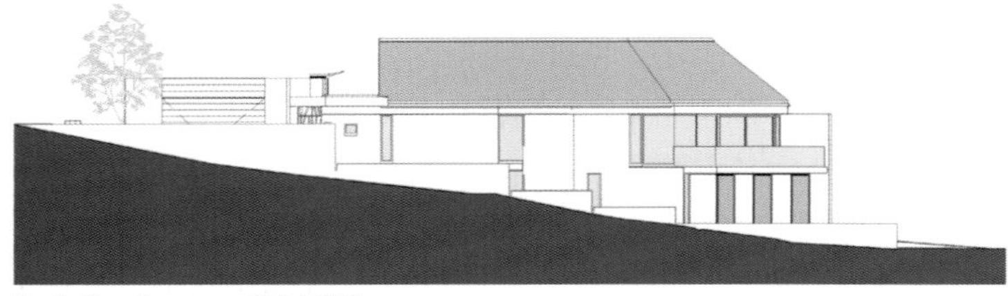

South Elevation 南面立面图

The design is largely formed by three contextual conditions. The first is the elevated entrance which placed the living spaces on the upper level and the bedrooms and playroom on the lower level. This decision allowed the living spaces to maximise the views of the bay and to see the water's edge. The second issue is the prevailing South-Easterly wind. The challenge here is that the views are in this direction and a large set of glazed sliding doors allows the maximum view. The third major issue is the sun on the North side. The response to this and to the South-Easterly wind was to position the pool in a courtyard on the Northern face that captures sun for the house and also creates a wind free outdoor space that can be enjoyed year round regardless of the wind.

The upper floor has been conceptualised as a single space holding the pool courtyard, an elevated entrance hall, a kitchen with a large table and a dining and living space on the East edge. The L-shaped space is broken up by level changes which create distinctly different spaces. A conical flue made from Corten steel forms a visual element around which the spaces pivot, the rusted surface reflecting the coastal environment.

The thatched roof is supported on a perimeter I-beam. The timber tie beams of the thatched roof are replaced by steel tie rods that allow the open volume of the roof to form the top of the space, the woven thatching grass and the latter forming a counterpoint to the granite floors.

A curved Corten 'hat' sits over the braai flue and watches over the pool courtyard, a playful companion for the Corten cone that sits on the other side of the courtyard. The North face of the courtyard is formed by a timber wall. On the lower level, the off-shutter concrete slabs form the ceiling. The slabs retain rust marks from the steel that was laid in the slab and the chalk markings of the contractor.

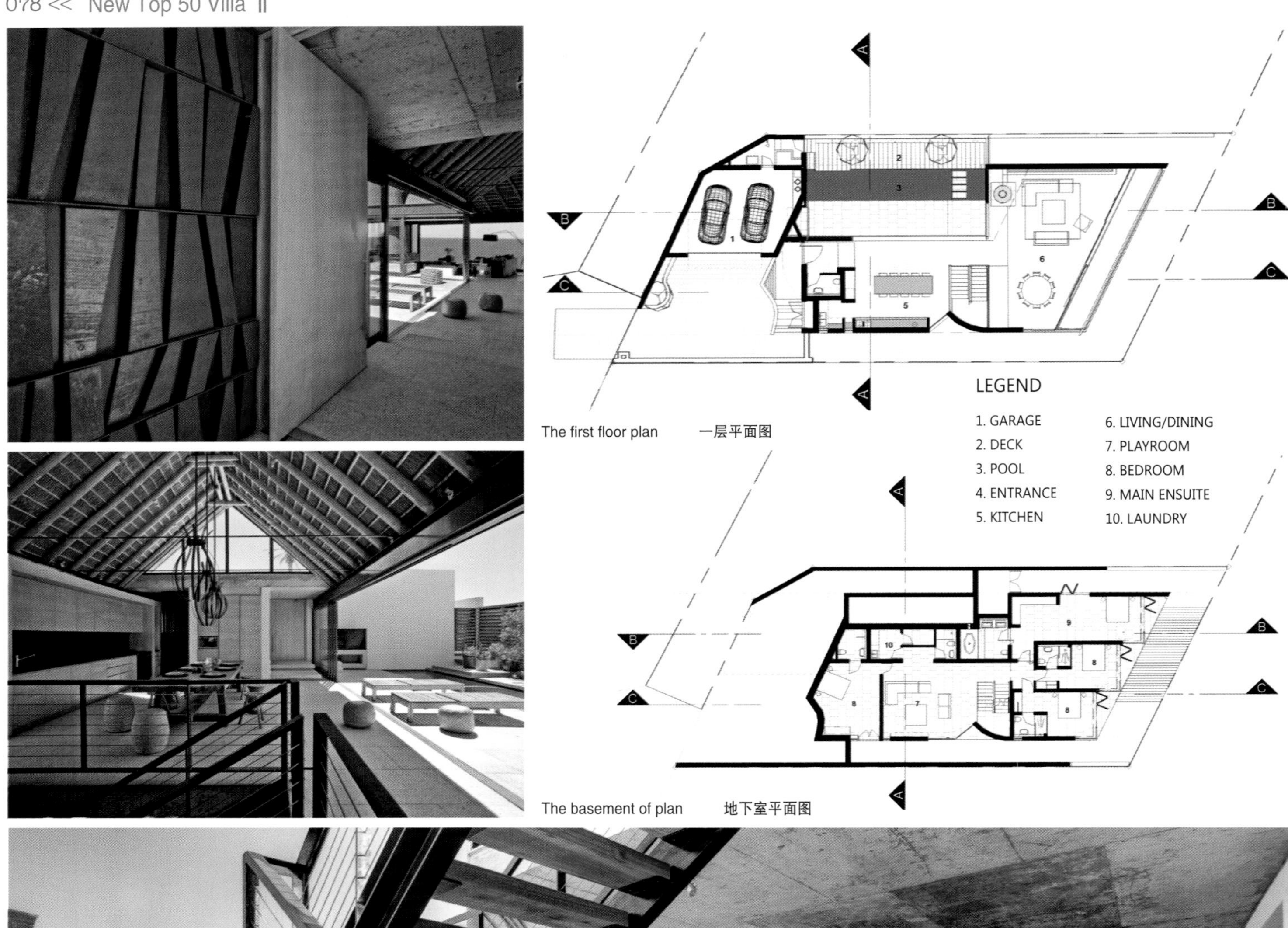

The first floor plan 一层平面图

The basement of plan 地下室平面图

The simple cellular bedrooms have a serrated façade that allows for corner glazing to maximise views. The curtain track was cast into the slab with a radiused corner that creates a cave-like space at night when the curtains are closed. Simple glass walls separate the en-suites from the bedrooms, allowing the two spaces to share a larger volume. The stair connecting the two levels is also made from I-beams with 75mm thick Eucalyptus planks forming the treads.

本案别墅位于南非西海岸雪莱。别墅所在地朝东，面向大海，将斯瓦特兰小镇后的群山风景尽收眼底。

本案的设计充分考虑了三个重要条件。首先是受坡度影响，房屋的入口被抬高，因此设计师将起居室安排在地面层，卧室和娱乐室则被安排在底层，这样就能使两层都拥有最佳的观海视野。其次，考虑到此地盛行东南风，设计师特别利用滑动玻璃门，既挡风又不影响观赏风景。第三，因为该案地处南半球，阳光自北而来，加之考虑到风向，设计师特别在北向设计了一个带泳池的庭院，以充分享受海风和阳光。

地面层空间被设计为一个独立空间，拥有带泳池的庭院和进门处的大厅，以及带有宽大餐桌的厨房和餐厅，客厅则安排在东边。该建筑布局自然形成 L 形，而餐厅和客厅就在 L 形的两端。客厅一侧有一个抢眼的锥形火炉通风口，耐候钢制表面已经被海洋气候腐蚀得锈迹斑斑。通风口顶如同戴了顶耐候钢制的帽子，俯视着院中的泳池。

颇有特色的茅草屋顶由工字钢结构支撑，地面则铺上了花岗石。庭院的北边，建起了一座木墙。而通往卧室层的楼梯也采用工字钢结构做框架，而卧室层的顶棚为混凝土，还特意保留了建筑时的斑驳印迹。卧室的玻璃为内部提供了最大的观海视野，窗帘的安装轨道则顺应转角的落地窗做成了圆弧形。

Villa G

G 公馆

- Design Agency: Saunders Architecture
- Designer: Todd Saunders
- Location: Bergen, Norway
- Area: 368m^2
- Photographer: Bent Ren é Synnvag, Jan Lillebo

- 设计单位：桑德斯建筑事务所
- 设计师：托德・桑德斯
- 项目地点：挪威，卑尔根
- 项目面积：368m^2
- 摄影：本特・勒内・辛瓦格，贾恩・里勒波

Architect Todd Saunders landed a creative and dynamic client when he got the job to design a house on the south-west coast of Bergen, Noway. The result is a rare and beautiful, and Saunders has said that the process has made him into a better architect.

Villa G lies like a white landmark in the soft landscape at Hjellestad, near Bergen. The house is large yet not dominating, modern but not pretentious. The house has a futuristic form but is built with traditional Nordic materials and architectural elements with a good basis in Norwegian building methods.

The wooden cladding on the house consist of 3 different size mounted in a random pattern. The house has an over built outside space and covered and the second floor covers the entrance below helping the house work together with the rough climate on the west coast of Norway.

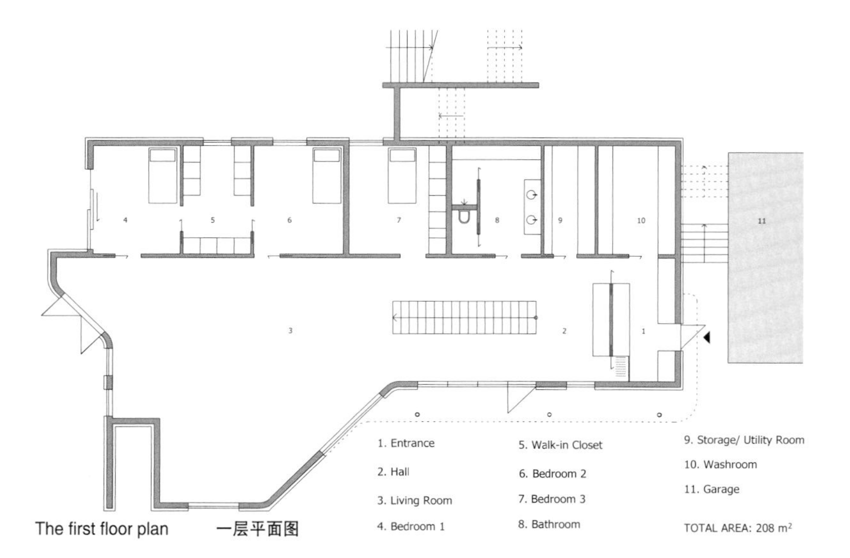

The first floor plan 一层平面图

The stair is one solid piece of 1cm thick steel, galvanised with white sand corn making it slipp resistant. The stair is produced locally, weighs almost a tonne, and had to be lifted into place by a crane through the window in the roof.

The client wanted a house with clean lines without any visual noise and clutter. This is one of the reasons that most of the closets and storage spaces are integrated into, so – called, thick wall – walls that are at least 60-70cm deep. The kitchen bench is 8 meters long and has plenty of drawers for kitchen equipment and even other things that need to be stored away. None of the electrical outlets are visible and all technology controlled by a main control panel in the kitchen. The client admits that he is a "gadget freak" and the house reflects this part of his personality.

设计师受一位创意十足的客户的委托为挪威卑尔根西南海岸的一座别墅进行设计。该别墅命名为 G 公馆，坐落于卑尔根附近的 Hjellestad 地区，是一座纯白的建筑。别墅很大，却不显浮夸，颇具现代特色，却无造作之气。尽管从外观上看，别墅新潮时尚，但其所用材料多为传统的北欧建筑材料，采用的建筑元素也具有挪威的建筑特色。

别墅外墙随机铺上了 3 种不同尺寸的木板。外部空间和墙壁厚度都比一般的住宅要大，且室外有楼梯能直通二楼……这些设计都是为了抵御挪威西南海岸的恶劣气候。室内楼梯使用了 1cm 厚的钢板，上面镀上了白砂粒，起到防滑的作用。该楼梯是本地出产，约重 1t，因此不得不使用起重机将其从屋顶的窗户吊进去。

业主希望房子拥有简洁的线条，且视觉上干净流畅。因此室内的橱柜等储藏空间都被设置在 60~70cm 厚的墙面里。8m 长的厨房流理台拥有许多实用的抽屉，用于厨具和小物件的收纳。厨房中见不到任何电路接口，且采用的所有科技均由一个主控制面板进行操控。业主自称“装置控”，而房子的很多设计也确实反映了他的这一特点。

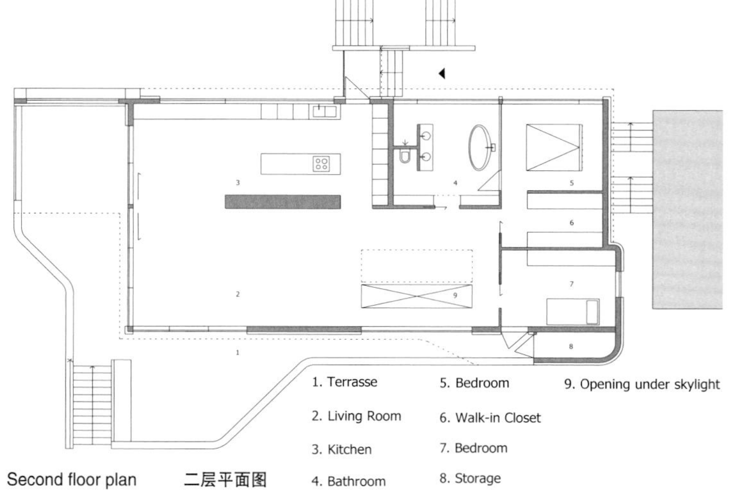

Second floor plan 二层平面图

1. Terrasse
2. Living Room
3. Kitchen
4. Bathroom
5. Bedroom
6. Walk-in Closet
7. Bedroom
8. Storage
9. Opening under skylight

Shenkeng Chen's House

深坑・陈公馆

- Design Agency: Enjoy-Design Engineering Co., Ltd.
- Designer: Xie Zongyi
- Location: Taiwan
- Area: 330m^2

- 设计单位 : 绝享设计工程有限公司
- 设计师 : 谢宗益
- 项目地点 : 台湾
- 项目面积 : 330m^2

Based on the original façade, the house is designed to be the Japanese style architecture that the client loves. The roof is partly tiled with black glass steel tiles, whilst the external wall is covered with stones, showing the house's classic aesthetic and simplicity. A light absorpting shade is used in the studio to welcome the sunshine outside into the interior space. The floor is covered with polished quartz bricks and it extends to the sunroom, which seems to magnify the space visually. Standing in the studio, you can enjoy the scenery of the Japanese style garden. The bedrooms are on the second floor and every suite has exclusive bathroom and lovely window scenery. The staircase and aisle is designed as modern area which is comfortable and practical in use. The original windows in the kitchen/ dining area are enlarged to connect the interior space to the green garden scenery. The kitchen cabinet coated with white paint uses glass boards, creating a bright and fresh atmosphere. The Tatami room with the door closed can be used as a single room while the floor is elevated to create a space of Japanese style.

在外观部分，以原来的建筑外墙为雏型，设计出屋主喜欢的日式建筑造型。除了在屋瓦部分利用黑色琉璃钢瓦为建材外，在墙面也敲掉原有砖墙及木作，改用抿石子的外墙设计，完美呈现古典美感、清丽、朴实的灰静身影。

画室利用采光罩让室内与户外阳光连接，加上地板整个翻新改用抛光石英砖，并将地砖延续至阳光屋，使空间有放大效果；而户外则因有抿石子墙、琉璃钢瓦屋顶等日式庭园风情，让作画累了的夫人可以饱览最爱的风景。二楼被规划为卧房区，在细部规划上，不但每间都是套房式规格，拥有专属浴室与对外窗景，另外在梯间与走道的设计上也不马虎，除了将原本残破的景象完全抹去，更加入现代舒适的视觉，展现出十足机能美感的个性山居。

餐厅与厨房和一的空间感更为通透开阔，并且将厨房改以白色烤漆与玻璃壁板设计，呈现出清丽的明亮气氛。将原有的两扇窗户尺寸放大，实现庭园餐厅的绿意与优闲感。关上门片即可自成一区的和室，透过架高地板的设计展现出屋主喜欢的日式恬淡风情；而为了让视野更开阔，特别将原本局限的窗型放大为落地窗，将满窗的绿意引进与女主人的画作相映成趣。

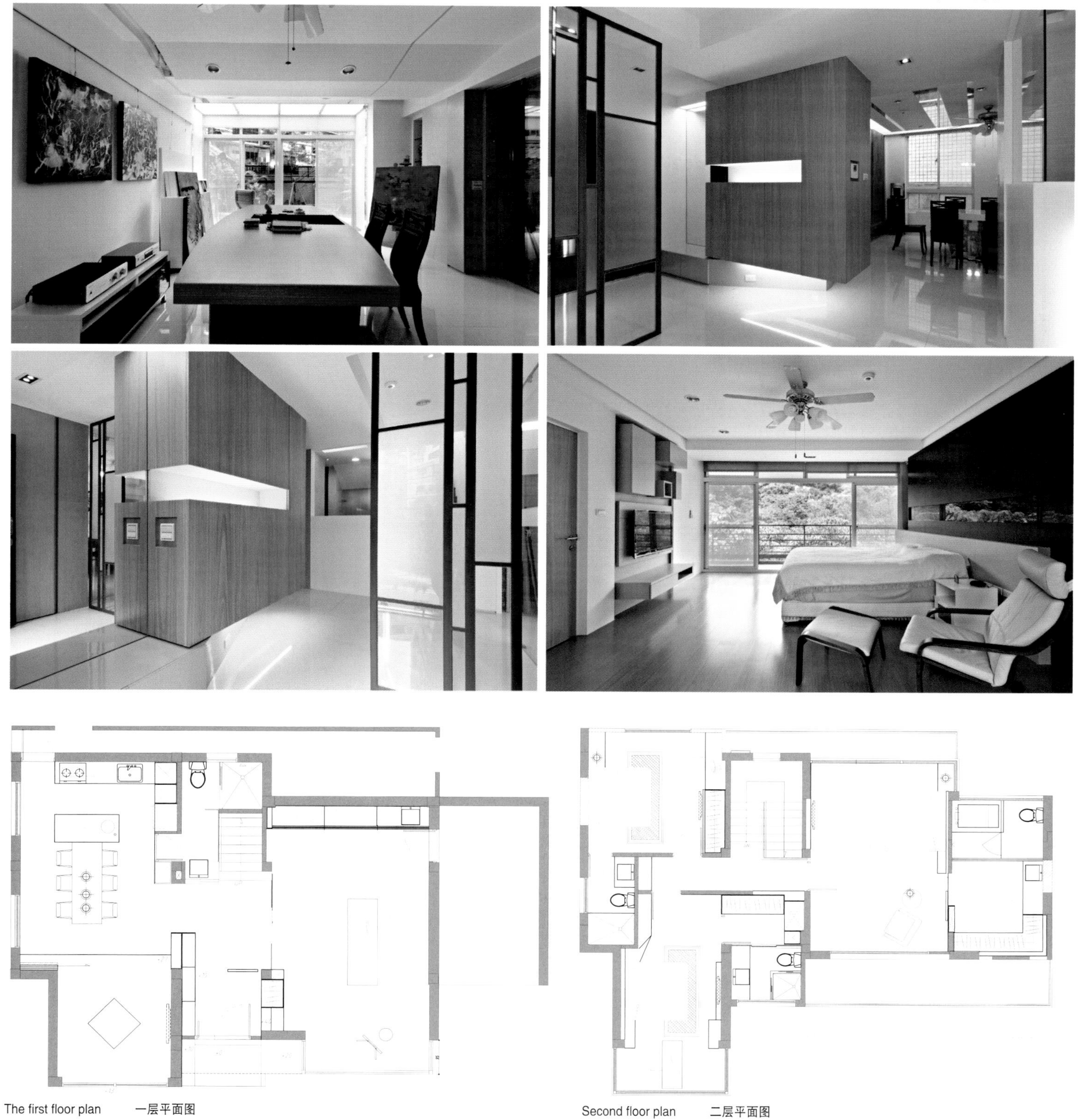

The first floor plan 一层平面图

Second floor plan 二层平面图

Taihu Lake Palace

苏州太湖天阕

- Design Agency: Shanghai Zhumu Space Design and Decoration Co., Ltd.
- Designer: Chen Jie
- Location: Suzhou, Jiangsu
- Area: 660m^2

- 设计单位：上海筑木空间设计装饰有限公司
- 设计师：陈洁
- 项目地点：江苏苏州
- 项目面积：660m^2

In the interior design, both the color of the fabric and paintings and the type of furniture emphasize the lightness.

The interior design has a modern country style. The first floor has the layout without any partitions, providing an uninterrupted view among the hallway, the living room, the piano area and the dinging room. A freestanding fireplace with two sides to fire not only decorates the space, but also improves the heating insulation of the house which has steel structure resulting in poor insulating effect.

The master suite is on the second floor. To keep the uninterrupted ceiling and skylight, the designer uses a half height fireplace wall to separate the bedroom and the bathroom. At the back of the fireplace, there is a double basin and a powder room. The pedestal toilet is separated by a partition wall.

The loft space is divided into two parts. One part is designed as a cantilevered balcony while the other is used as a living space and the hostess' office.

The basement is designed as a play room where you can watch movies, sing karaoke, enjoy wine in the cellar, play cards with family, play table tennis and take a bath or a foot bath in a spa area.

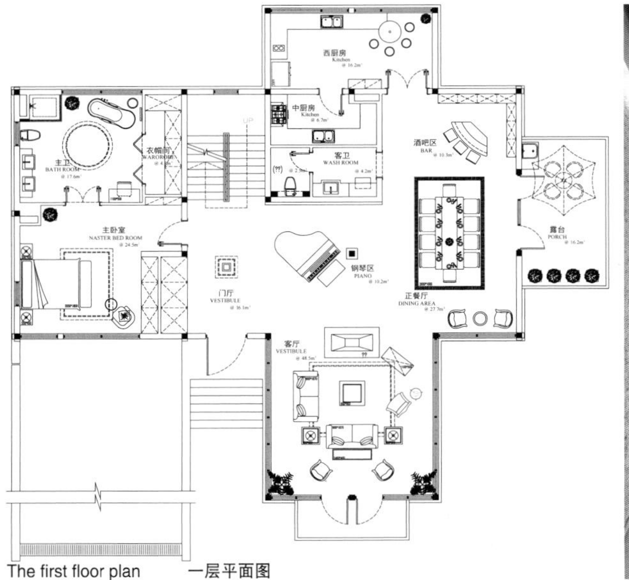

The first floor plan　一层平面图

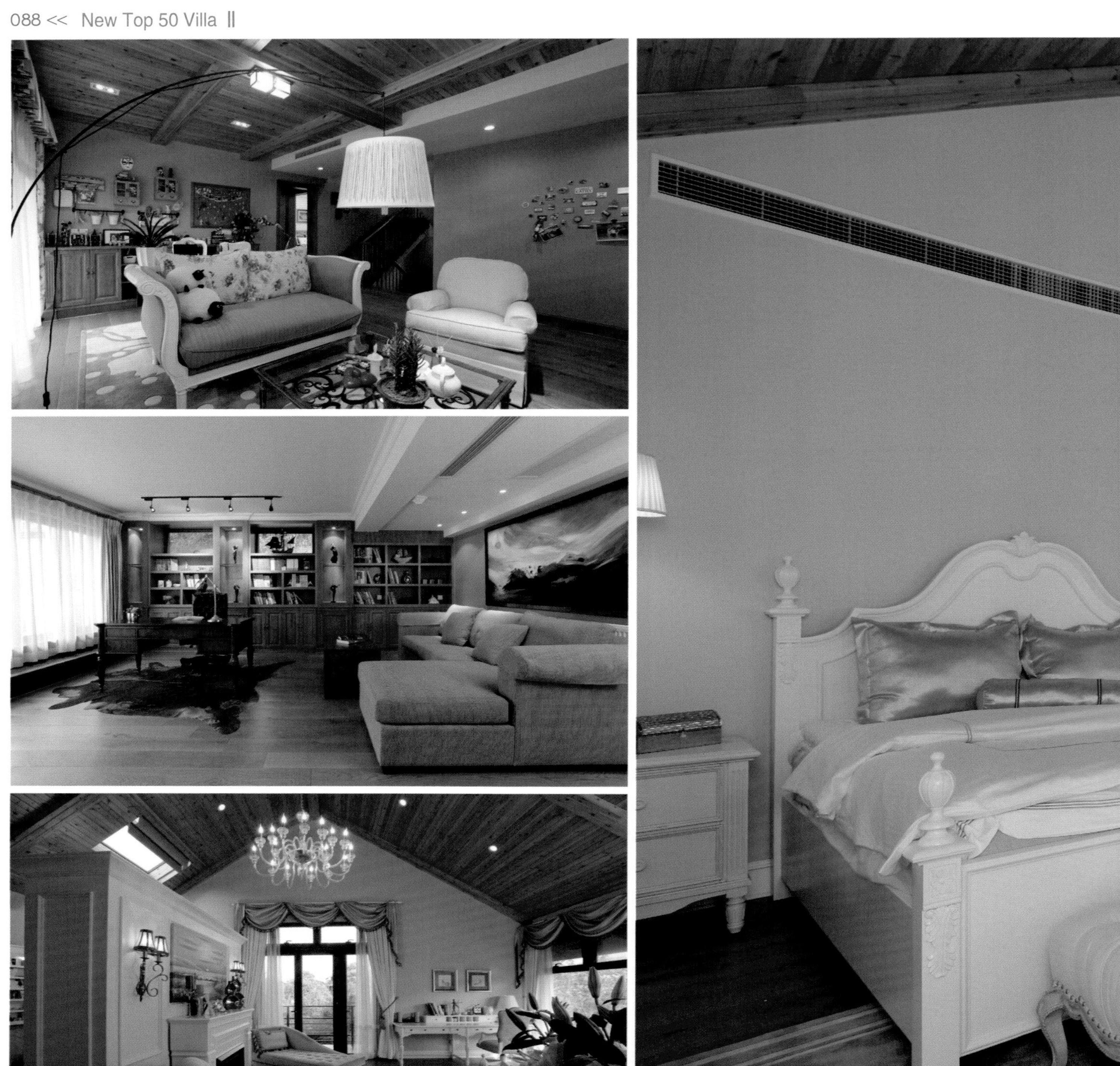

本案的整体设计，无论布艺色彩、家具款式、油漆色泽，都以轻盈为主。

室内采用了现代乡村的设计风格。一楼采用无隔断的布局，门厅、客厅、钢琴区、餐厅，一览无余。而一个独立的双面烧火壁炉，既起到美观的作用，又为这一钢结构、保温效果略差的屋子，在冬天的地暖采暖起到很好的补充作用。

二楼的主人套房，为保留顶棚的不间断及天窗的效果，不设计独立的卫浴空间，以一个半高的壁炉墙，将卧室与卫生间巧妙分隔，一边是壁炉造型，背后则设计成双台盆及化妆区。坐厕用一段由淋浴房的隔断延伸出来的矮墙巧妙隔开。

阁楼的空间被一分为二。一部分空间，可由二楼主人套房的更衣室内一部小巧的旋转楼梯走入。这部分阁楼，将与卧室相通的墙体打通，并略微外挑形成一个小阳台。另一部分空间，作为全家的休息空间，也成了女主人的小型工作室。

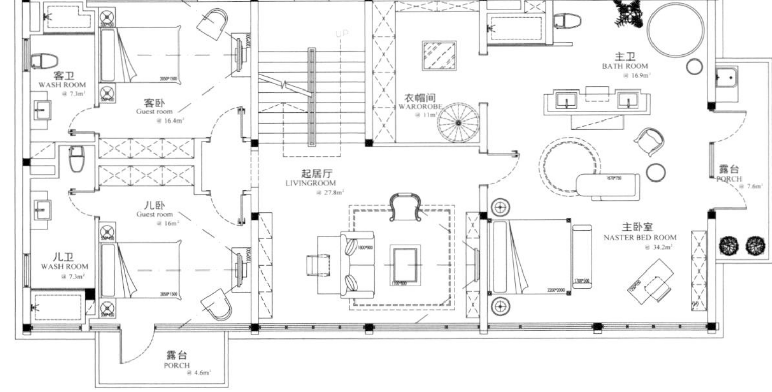

Second floor plan　　二层平面图

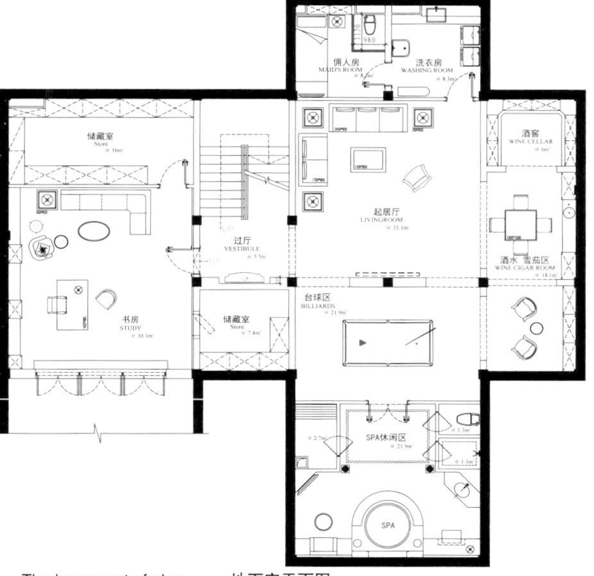

The basement of plan　　地下室平面图

4MS

4MS 别墅

- Design Agency: Pardini Hall Architecture
- Designer: Elisa Pardini
- Location: London，England
- Area: 255m^2
- Photographer: Carlo Carossio

- 设计单位：Pardini Hall 建筑事务所
- 设计师：艾莉莎・帕蒂尼
- 项目地点：英国，伦敦
- 项目面积：255m^2
- 摄影：卡洛・卡罗希欧

The client wanted to fully refurbish his 5 storey house and to extend it. The existing dwelling has a deep and narrow floor plate, with daylight penetrating from the east and west.

The proposed project aims to enhance the amount of natural light by investigating levels of transparency both vertical and horizontally, through a play of reflections and perspectives. That is the reason why we choose to build an extension in complete glass, and we used stainless steel for the structure in order to have the reflection of the green around. The depth of the building should generate a visual connectivity between the living space and the external landscaping, whilst maximizing transparency and natural daylight.

The choice Open space, creating a sequence of distinct functional areas where it is dismantled the concept of "room" for a context of spatial fluidity. The service areas including toilet and laundry cupboard are all together. The dining area is planned in the basement, along with the kitchen area, while the ground floor is dedicated to the living room and a library room. The creation of a core services behind the kitchen frees up the stairs by expanding existing wardrobes across wide spaces and create a space for living adjacent to the well-being. The bedrooms, on the three floors below, want to maximize the spaces with furniture bespoke.

Section　剖面图

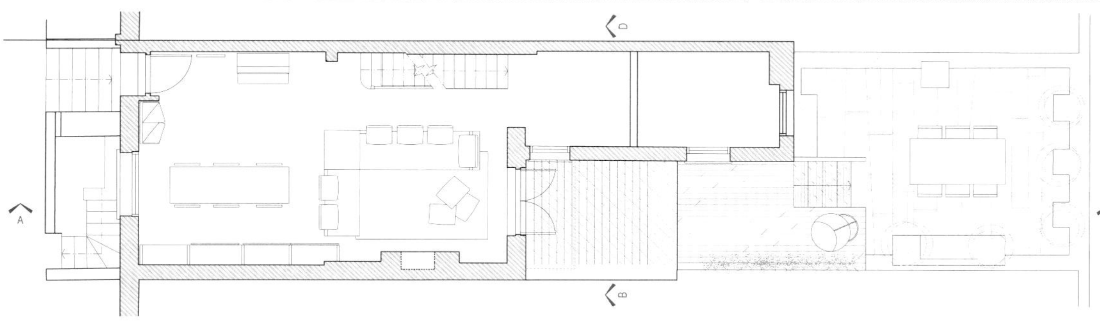

The first floor plan 一层平面图

业主要求对这栋 5 层小楼进行全面的翻新和扩展。因原楼的楼板厚而窄，光照只能从东边和西边微微透进来。所以该案设计的主要目的是通过对屋子各方面透光度的分析，从反射和透视的角度入手，加强屋子的采光。所以设计师在扩展空间时全面使用了玻璃，同时为了映照周围的绿色植物，在结构上还使用了反光的不锈钢材料。这样一来，房屋的空间感得到加深，室内空间和室外景观得以沟通，同时还最大限度地增强了房屋的透光性，使整个室内获得最佳的采光条件。

该案选择开放空间的设计，为改善空间流动性，将房间的概念重新定义，从而产生了一系列功能区域。其中，服务区域包括卫浴和洗衣房。餐厅和厨房设置在底层，而一楼则用作客厅和阅览室。厨房后的服务层，通过扩大原有的橱柜，以腾出楼梯的空间，从而形成一个休息室。分布在另外三层的卧室，则采用了特别定制的家具，来达到扩展空间的效果。

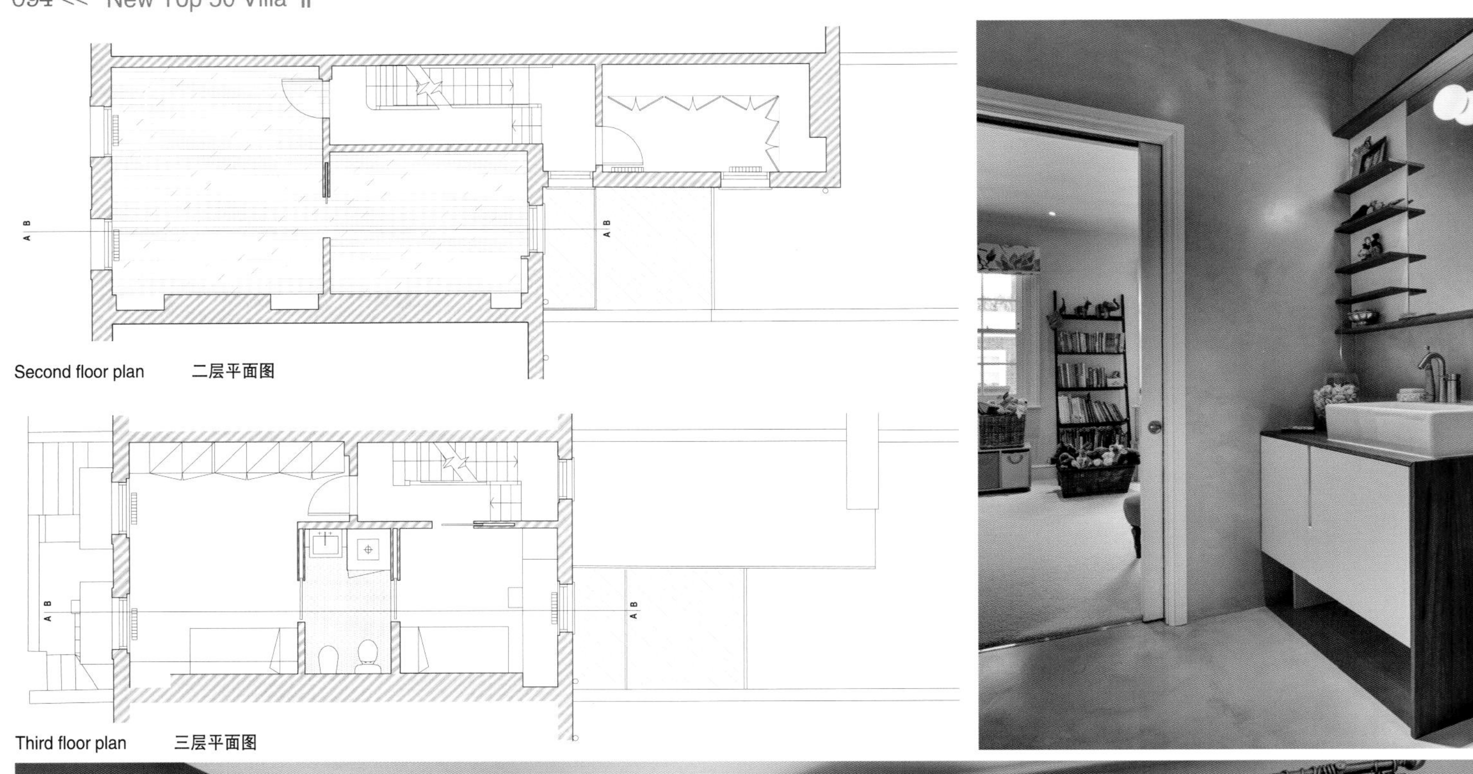

Second floor plan 二层平面图

Third floor plan 三层平面图

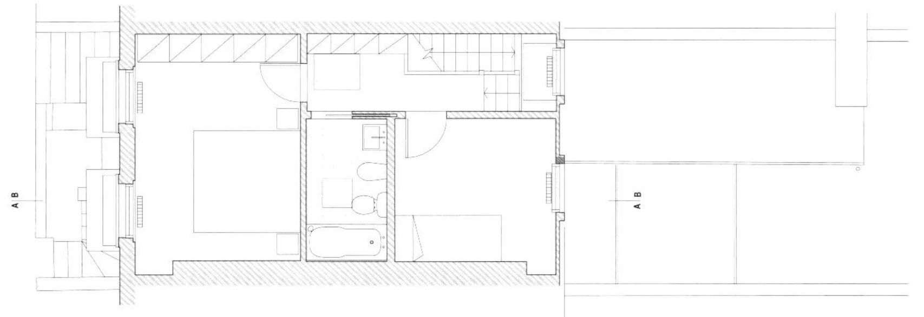

Four floor plan　　四层平面图

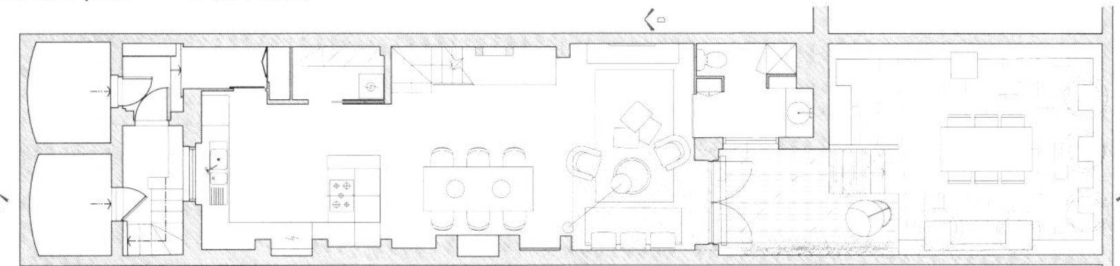

The basement of plan　　地下室平面图

Dazzling in Golden Blackberry Rose

金醉荼蘼

- Design Agency: Pinki Interior Design Liu& Associates IARI Interior Design
- Designer: Danfu Liu
- Location: Jinan, Shandong
- Area: 370m^2

- 设计单位：PINKI（品伊国际创意）& PINKI DESIGN 美国 IARI 刘卫军设计事务所
- 设计师：刘卫军
- 项目地点：山东济南
- 项目面积：370m^2

The first floor plan 一层平面图

Blackberry rose is a flower that can make a Buddha smile. However, it always used in the ancient poetry to express sadness. Blackberry rose, a white warm-season and sunny flower, is the last one to bloom in the spring, most of which are not blooming until the summer. Gold is a stable element which does not easily react to any other elements. Thus gold blackberry rose is used as the theme to express unique personality and maturity after experiencing every story, which seems to be successful but lonely. Actually, it refers to some kind of insights after you reach a peak of your life.

This project is designed to be golden-white tone. The living room, kitchen, master room, main bathroom and furniture displayed in this house, even small decorations are mostly golden and white; with the support of the dark-colored furniture and various lines or circles, mature quality can be shown even from the very detail corner. Colors like brown, coffee and black help to enhance the dimension sense, to make this huge space more delicate and dignified with outstanding nobility. This space does not only emphasize luxury and gorgeousness, but also tell stories about clear peace. And only when you are settling yourself down without feeling any anxiety or fickleness, can you feel it. Calm down, and then you can meet the most amazing part of this design.

荼蘼，别名佛见笑，在古今文人的诗词歌赋中的寓意几乎都是伤感的。荼蘼是春天最后开的一种白色、喜温向阳的繁花，往往都是春天过后，直到夏天才开花。稳定的金元素不容易与其他元素产生化学反应，而金醉荼蘼这个主题主要表达的是一份有经历与故事的成熟，不脱离主流价值观的独特气质，看似有所成就而孤独，本质上是寓示人生修行至高点的一种顿悟。

本案以金色、白色为设计主调性，不论是客厅、厨房、主卧、衣帽间、主卫还是到陈设配饰的家具、饰品都使用了大量的金色和白色，加以深色的家具沉淀空间，粗细不同、繁简有异的直线和曲线，在细微处渲染出成熟的气质。咖色、棕色、黑色等浓重的色调增加了体量感，使高大空阔的空间更具深宅高院的沉稳凝重，贵族气息不言而喻。整个空间不仅仅表达了奢华与高雅，静心感受，不焦虑、不浮躁，才可悟到明朗安详，你懂便能懂。

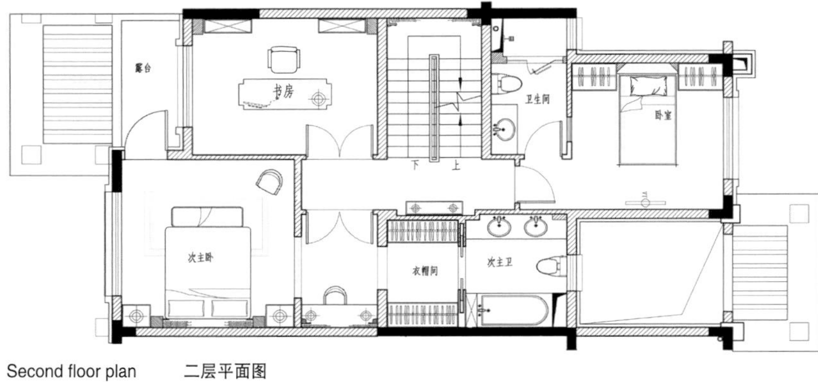

Second floor plan　　二层平面图

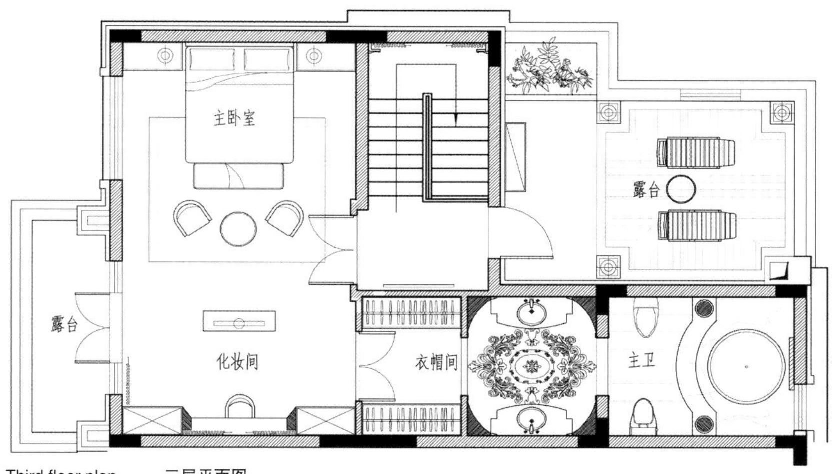

Third floor plan　　三层平面图

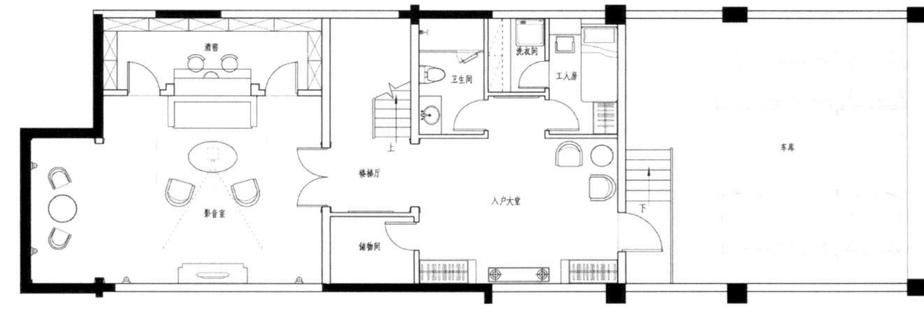

The basement of plan　　地下室平面图

Gorgeous Manor

华府庄园

- Design Agency: Wuxi Dong Yi Ri Sheng Decorative Engineering Co., Ltd.
- Designer: Wei Qun
- Location: Wuxi, Jiangsu
- Area: 350m^2

- 设计单位：无锡东易日盛装饰公司
- 设计师：韦群
- 项目地点：江苏无锡
- 项目面积：350m^2

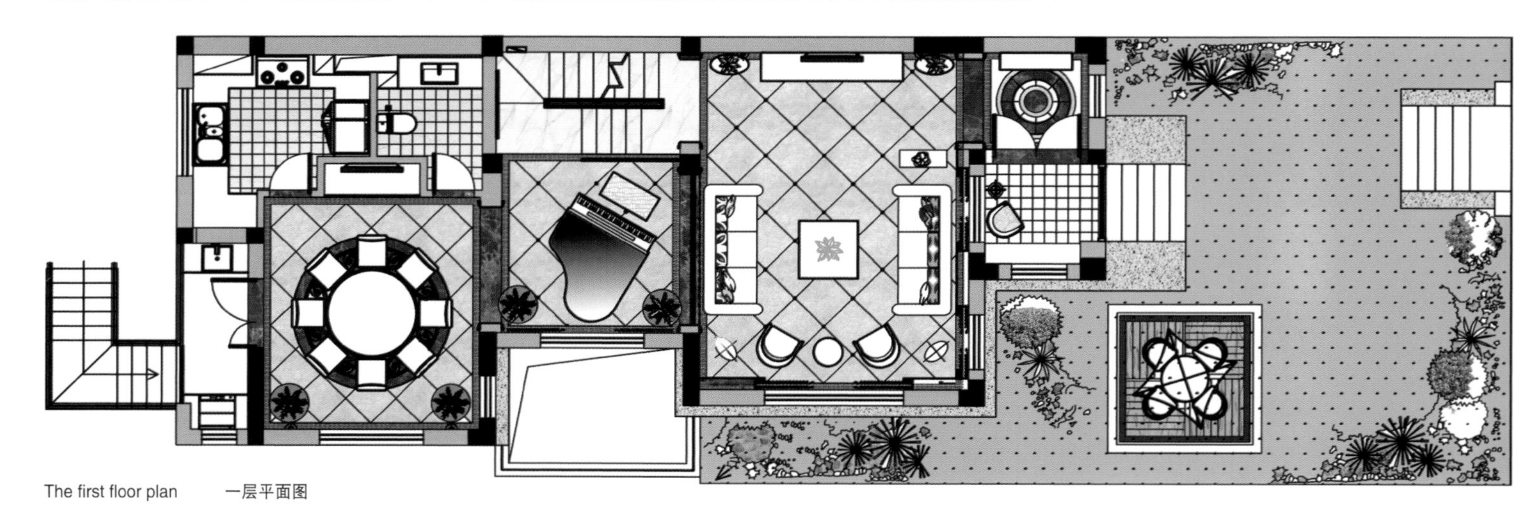

The first floor plan　　一层平面图

This villa sits in a perfect landscape system which enjoys rich natural resources. It is mixed into a home with particular features where you can see classic European elements such as French and Italian styles when you are attracted by Chinese features in medieval of Qing dynasty. The space is filled with endless noble atmosphere because of the rich and exquisite details.

The Rococo sofa in the living room gives people a sense of comfort and softness, whilst the natural lights coming from the large floor windows help to create a wonderful environment to live in. In the dining room, space structure remains open, and it creates a comfortable space feelings and luxurious atmosphere of European palace by using elegant floor bricks and wallpapers. These exquisite wallpapers and delicate wall fittings in bedroom give off rich French cultural fragrance. Cleverish decorations beside the beds also help to add some steady visual perception. Space in bathroom is much enjoyable by combining the mosaics and pattern wall, showing much more luxury sense.

此套别墅拥有完美的自然资源和景观体系。本案业主对古典欧式有着浓烈的喜爱情节，我们通过与原建筑外观以及对庭院和别墅周边的整理，再结合业主提出的意见，最终将本次方案定位为法式新古典风格。

客厅洛可可风的沙发有着细腻且柔适的触感，给主人最惬意的居家感受，大面积的落地窗给予空间最大面积的自然采光，椭圆的顶棚设计嵌于笔直的空间线条内，方中有圆的造型使得空间肌理更为细致。餐区内保持开敞式的空间结构，通过地砖及壁纸的运用，营造出欧式设计的豪华、舒适的空间氛围，金色和米色的运用，更是将空间氛围升华，尽显贵气十足的空间气质。卧室精致的壁纸，别致的壁灯，颇显厚重的法国文化气息，小巧的床边装饰物，给人沉稳的视觉感受。主卧内宽敞而明亮，通过布艺软包形成的床背景墙，凸显出主卧的尊贵性，设计师于现代古典的奢华之中融入现代的流畅之美，迸发出强有力的个性设计。整个卫浴空间令人心旷神怡，马赛克拼花墙面结合，更显奢华。

Schumann

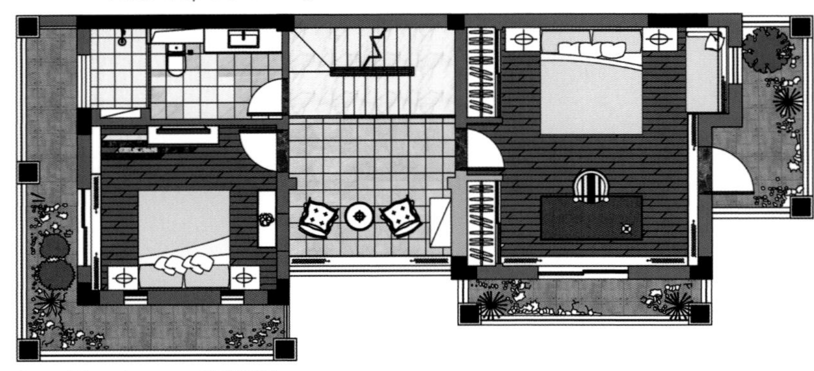

Second floor plan 二层平面图

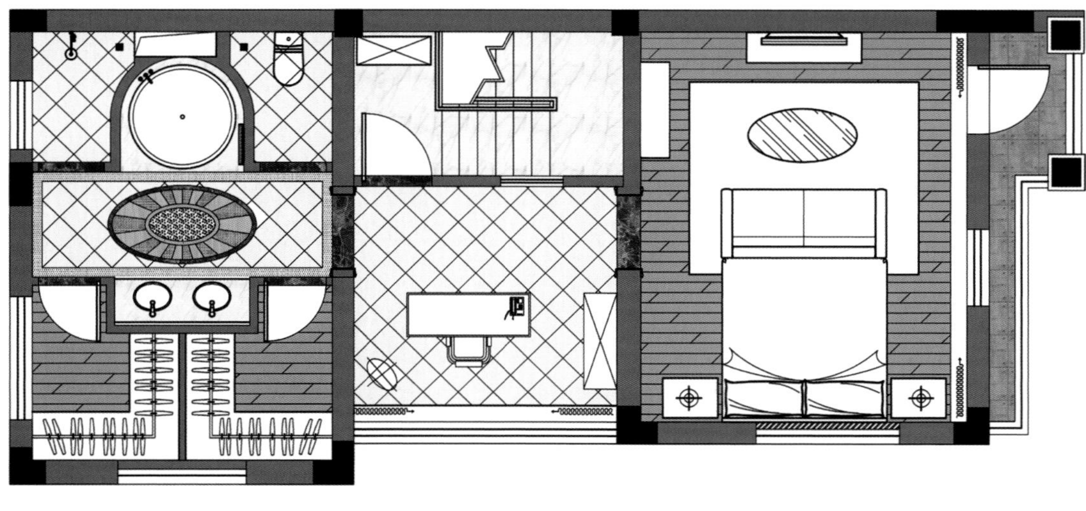

Third floor plan 三层平面图

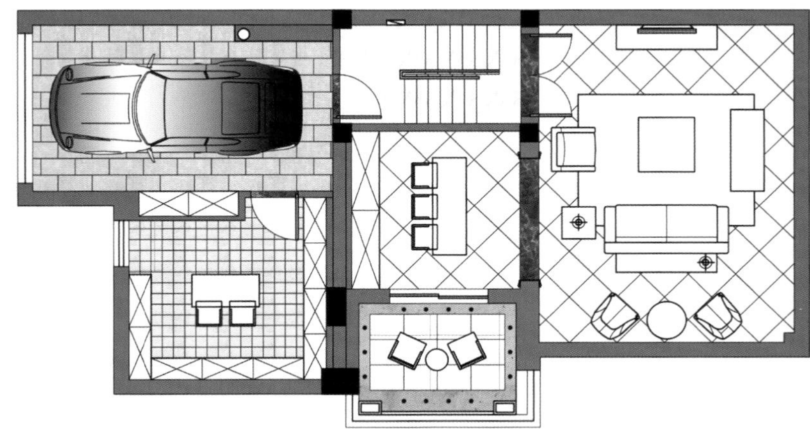

The basement of plan　　地下室平面图

Aesthetic Residence

唯美・释意

- Design Agency: ZhongCe Decoration
- Designer: Duan Qiyan
- Location: Kunming, Yunnan
- Area: 260m^2

- 设计单位：中策装饰
- 设计师：段其艳
- 项目地点：云南昆明
- 项目面积：260m^2

The Southest Asian style residence show its aesthetic value in every detail of furniture and accessosy. With a tranquil and elegant environment and beautiful scenery, many cultural elements are used in the design to enhance the space. Colors like green and yellow are adopted, along with metal crafts and handmade cushions, to add richness to the space from aspects of style, color and form.

Wherever you are in the room, you can enjoy a relaxing and comfort atmosphere as natural as the afternoon sunshine.

本案是典型的东南亚设计风格，设计师从设计到陈设配饰都给以唯美的释意。由于项目环境幽雅，自然景色绝佳，而东南亚风格又属于原味朴实的设计手法，所以在陈设配饰方面，给予空间大量的自然元素。一点新绿，一点藤黄，一尊金银铜器，一款极具手工艺味的靠垫，从风格、色彩和形态上给其丰盈而多变的表达。

不论你处在房间的什么位置，都能感受自然闲适的生活韵味，如慵懒随意的午后阳光，散发着迷人的自然、健康气息，搭配别具特色的配饰，整个房间温润且有趣味，让人无法自拔。

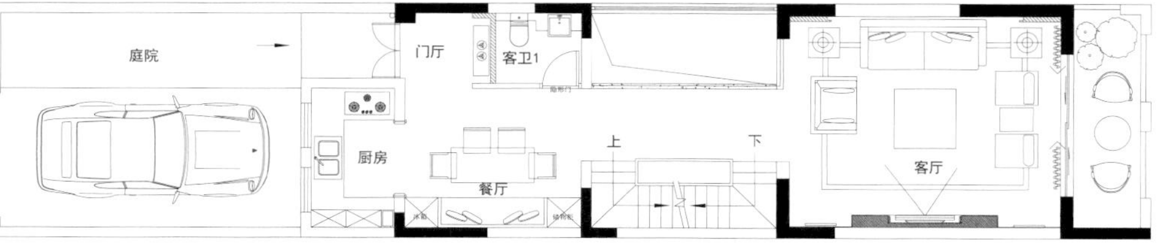

The first floor plan　一层平面图

Second floor plan 二层平面图

Third floor plan　　三层平面图

American Charm

美式风情

- Design Agency: Zhongce Decoration
- Designer: Wang Xinyong
- Location: Yunnan
- Area: 360m^2

- 设计单位：中策装饰
- 设计师：王新勇
- 项目地点：云南
- 项目面积：360m^2

This project displays the American style aesthetic, an mixture of various spatial elements.

The living room has clean and sprightly lines. Emphysizing on the design of the fireplace and the use of handmade ornaments, things like antique floor bricks and wallpapers, fabric sofa made of wood, handmade carpet and antique art decorations are used in the interior design to create an American style space which is spacious and historical. The open kitchen is typical of American style: the practical dining table, the spacious kitchen ranch which can include a double-door refrigerator, the antique bricks and wood cabinet, etc. here are showing the American charm.

This space is free and relaxing, simple and nostalgic, practical and comfort, which proves that the American style is not only a style, but also a living attitude.

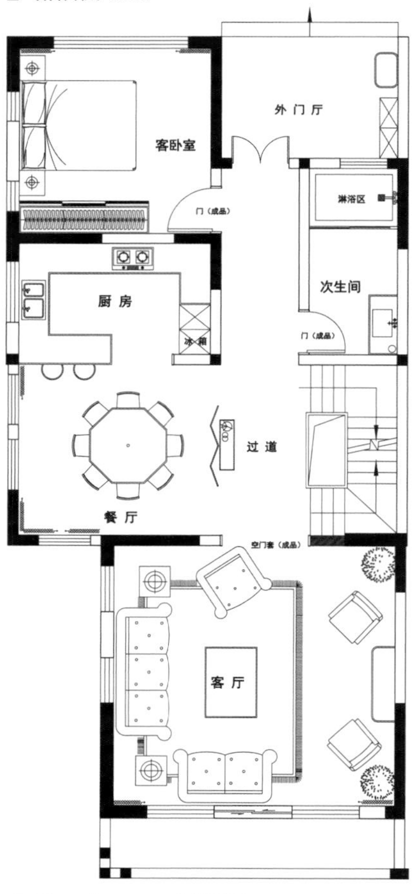

The first floor plan　一层平面图

本案非常完美地体现了以宽大、舒适、杂糅各种风格元素而著称的美式空间美学。

客厅的线条简洁明快。壁炉与手工装饰，仿古地砖和墙纸，加上木制布艺沙发、手工地毯，以及仿古艺术品的软装摆设，营造出了宽敞而富有历史气息的美式风格客厅。敞开式厨房设计，让人错以为走进了美国人的厨房：方便实用的便餐台，可以容纳双开门冰箱的宽敞位置和足够的操作台面。仿古面的墙砖、实木橱柜……无一不在散发着浓郁的美式风情。

自由随意、简洁怀旧、实用舒适，美式居家说的不只是一种风格，也是一种生活态度。

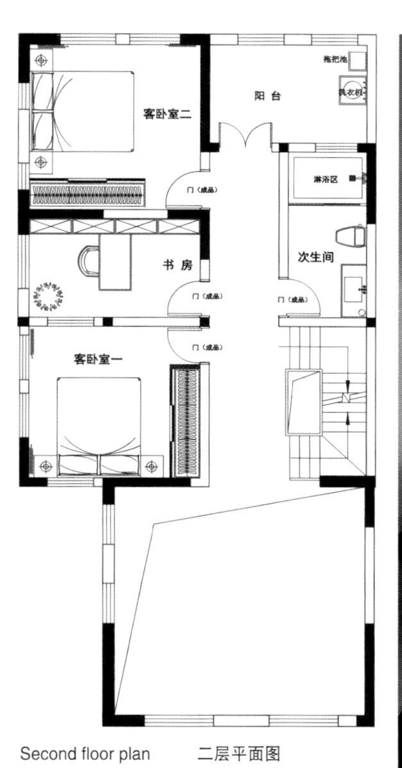

Second floor plan 二层平面图

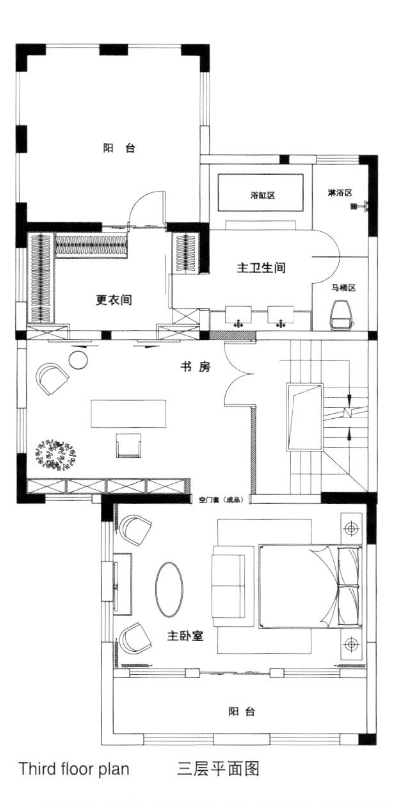

Third floor plan 三层平面图

酒吧区
保姆间
淋浴区
门（成品）
卫生间
门（成品）
多功能间
门（成品）
门（成品）
车 库

The basement of plan 地下室平面图

Chinese Spirit

中式风骨

- Design Agency: Zhongce Decoration
- Location: Yunnan
- Area: 360m^2

- 设计单位：中策装饰
- 项目地点：云南
- 项目面积：360m^2

The real Chinese modern style not only shows Chinese traditional charm, but also have modern features.

In this project, Chinese cultural elements integrate well with modern fashionable elements, to fully express the designer's deep understanding of Chinese culture and contemporary fashion.

The interior design reflects a kind of living style, moreover, it is also a living attitude. Modern Chinese style design is the way for Chinese people to inherit the culture of China, the greatest country with the world longest history. Also it is a bright star in the sky of interior design styles!

The first floor plan　一层平面图

真正的现代中式风格，是既能体现中国传统神韵，又具备现代感的设计。本案中，各种中国文化元素与现代时尚元素交相呼应，互为表里，相互碰撞，又相互衬托，既相得益彰又水乳交融。本案的设计将设计者对传统中国文化的了解及对当代社会时尚元素的把握展现得淋漓尽致。

室内设计，反映的其实是一种生活方式，是对生活的一种态度。现代中式风格设计，是现代中国人传承自己这个世界上历史最悠久、最古老伟大民族文化的方式，也成为了室内设计风格中一道亮丽的风景！

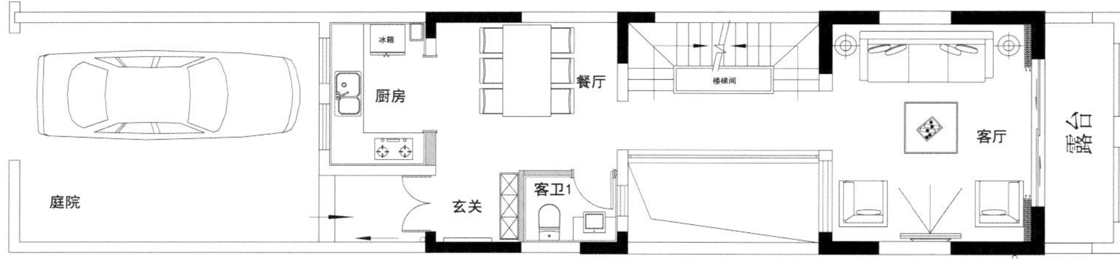

Second floor plan 二层平面图

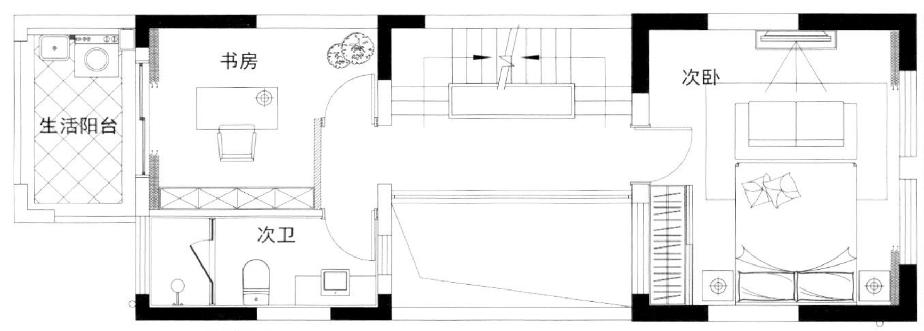

Third floor plan　三层平面图

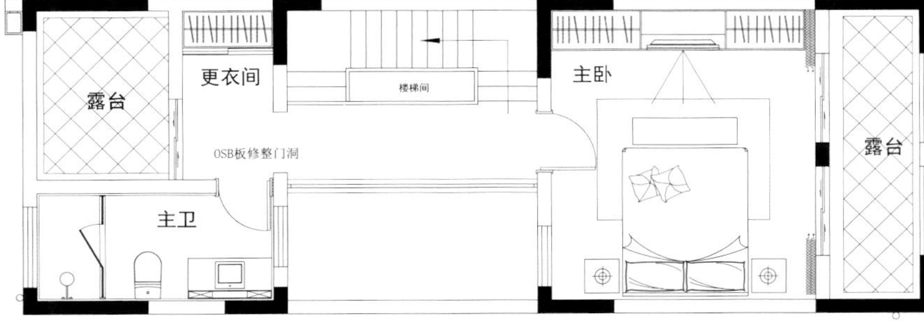

Fourth floor plan　四层平面图

Antai Villa

安泰别业

- Design Agency：Dashu Luxuries– You Weizhuang Design
- Designer: You Weizhuang
- Location: Nantong, Jiangsu
- Area: 400m^2

- 设计单位：大墅尚品—由伟壮设计
- 设计师：由伟壮
- 项目地点：江苏南通
- 项目面积：400m^2

The house can grow up with its owners. In different time, the house has different expressions and feelings of life. Thus, the project is designed with the creative concept of enjoying life, to create a space full of flavour of life. It breaks the rules of fixed design style to enable the space to show different faces as the life experiences are accumulated. The main palette is wood, supplemented by camel, creating an elegant and luxury space. Quiet decoration lines shines or disappears in the design, adding classic charm to the space.

The first floor plan 一层平面图

“家应该可以与主人一起成长”。家在不同时期可以有不同的呈现，展现不同的生活情怀。为此，这套住宅设计以“享生活”为主题创意概念，让家的空间从装修到家具，从装饰到陈列，无不展现出“生活”的浓厚气息，打破了空间固有设计风格的束缚，让空间可以随着生活阅历的积淀而呈现不同的变化。

在规律中进行变化，展现使用者本身的心境故事。整个空间以沉稳的木色为主基调，驼色为辅，空间环境雅致、奢华。宁静的装饰线条，时而闪烁耀眼，时而隐藏其型，或灵或透，一切都是最自然的呈现，让空间弥漫高贵的古典韵味。

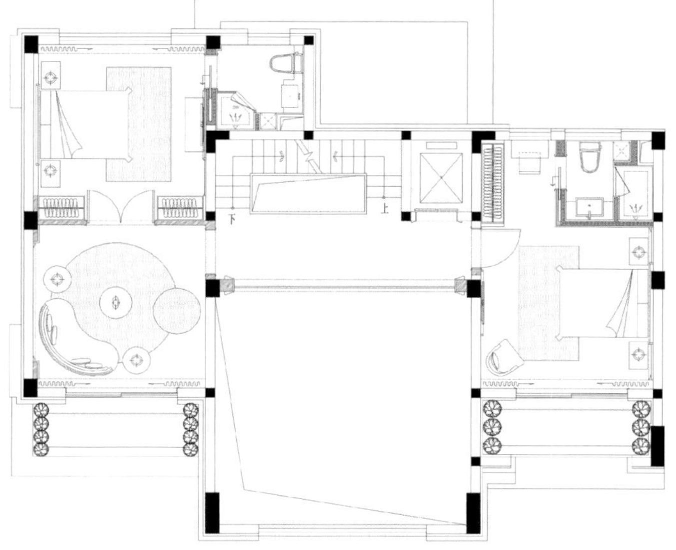

Second floor plan　　二层平面图

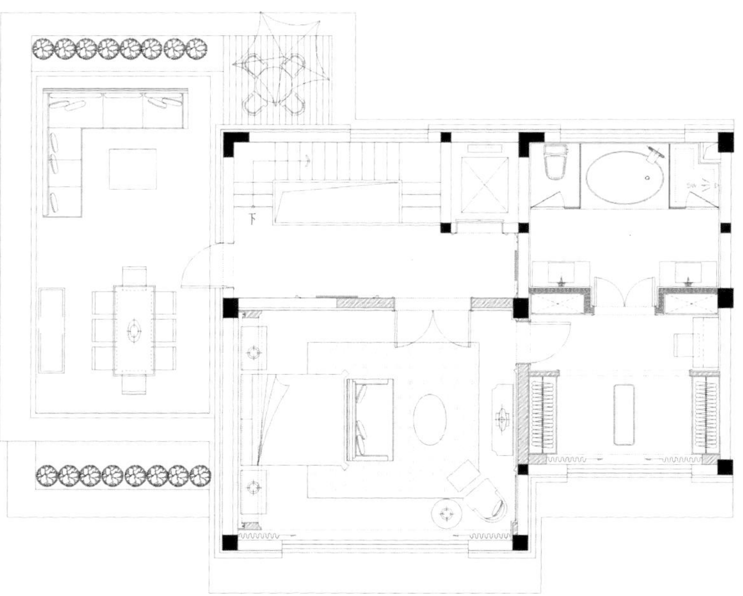

Third floor plan 三层平面图

The basement of plan 地下室平面图

Ba De Zhang's House

八德张宅

- Design Agency: Jiintorng Home Decorating Studio
- Designer: Zhang Renchuan, Deng Weiru
- Location: Taiwan
- Area: 990m^2

- 设计单位：堇桐室内设计
- 设计师：张仁川，邓维如
- 项目地点：台湾
- 项目面积：990m^2

It is a two-storey villa with a big front yard filled with many art collections. As the beautiful lighting design echoes with the stars at night, the exterior space is well enhanced to add more elegance to the villa. Stepping into the hallway, you will be welcomed by the humanistic art atmosphere.

Given that the client's occupation, the designer use various special materials. For instance, the unavoidable cylinder in the living room is covered with shell-like bricks, on which the sunshine can reflect different lights from different directions. Looking through the glass window of both sides, the great green scenery can be appreciated. While the timber stair of steel structure leads the green scenery of the atrium into the interior space. The single space of the kitchen is separated into two areas including kitchen and dining area, in which people can enjoy a good view of the green scenery and the stars at night through the big window from the ceiling to the ground. A unique space is used to exhibit the owner's collection of tea-things.

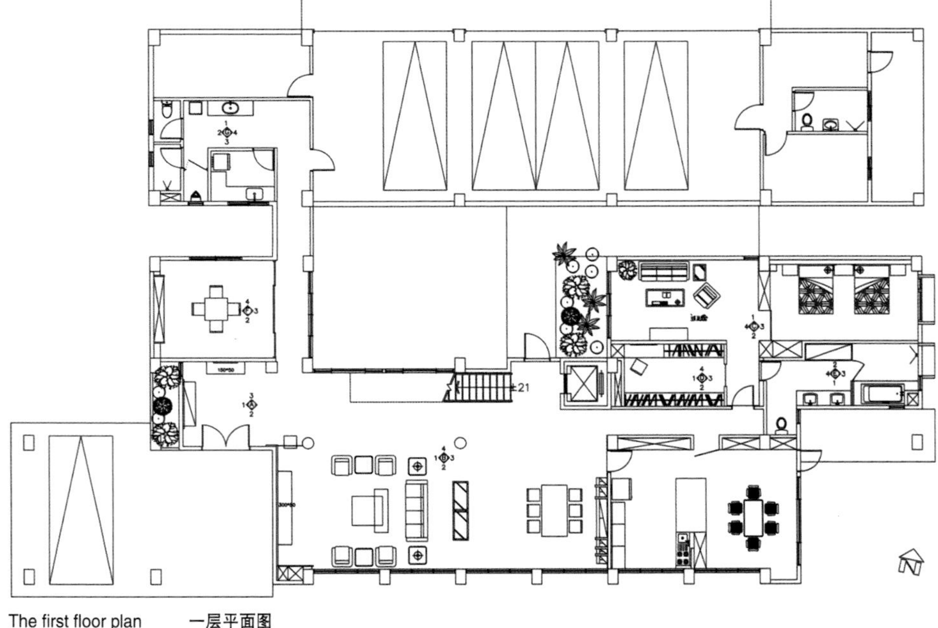

The first floor plan　　一层平面图

此案共有两个楼层。除了开阔的前庭视野，还将许多艺术收藏品融于空间动线里；室外更结合灯光设计让建筑物在星空的洗礼下闪闪发光，实为一个拥有完美景观庭院的别墅。

Second floor plan 二层平面图

踏入门厅，便能感受到一股浓浓的艺术人文气息。由于业主的企业性质，设计师特别融入许多特殊面材，例如客厅中不可避免的圆柱，使用了特殊的贝壳砖，并发挥它独特的性质，在不同的角度随着阳光走向可折射出不同的灿烂光泽。除了前景双面景观绿地之外，还规划了穿透性十足的钢结构实木楼梯，让视野得以延伸至中庭的绿意，达到处处皆是景的意境。独立空间的厨房与餐厅为一体，落地大窗让绿意和星光也能进入餐厅，让家人共享美好餐点时光。开放的端景区摆放着业主收藏的各式茶器，随着季节的不同可搭配不同的茶、器、花、画、琴进而转变茶席的氛围，置身其中静静赏玩。

Guanxi Chen's House

关西陈宅

- Design Agency: Jiintorng Home Decorating Studio
- Designer: Zhang Renchuan, Deng Weiru
- Location: Taiwan
- Area: 712m^2

- 设计单位：墐桐室内设计
- 设计师：张仁川，邓维如
- 项目地点：台湾
- 项目面积：712m^2

In this case, two key conditions have been taken into consideration: the window lighting and the wind direction. There is a home lift to connect three levels of the house. Each level has its functions and privacy. On the first floor, there are secluded rooms for parents and servants, as well as the dining room of the neoclassic style. In the dining room, the floor and the wall are decorated with cutting stone, whilst the European style furniture creates a modest luxury atmosphere which fit the client's personality. Different from the first floor, the interior design of the second floor emphasizes the modern concise style that is simple but elegant. Various materials of different qualities are used in the black-and-white wall to show its black palette, which adds modest and rich temperament to the space. The third floor includes a home-theater and a tea room. The tea room is divided into two parts, including an open area and a Tatami room, which can be also used as the guest room. The palette of the Tatami room is silver white, creating a quiet and elegant atmosphere.

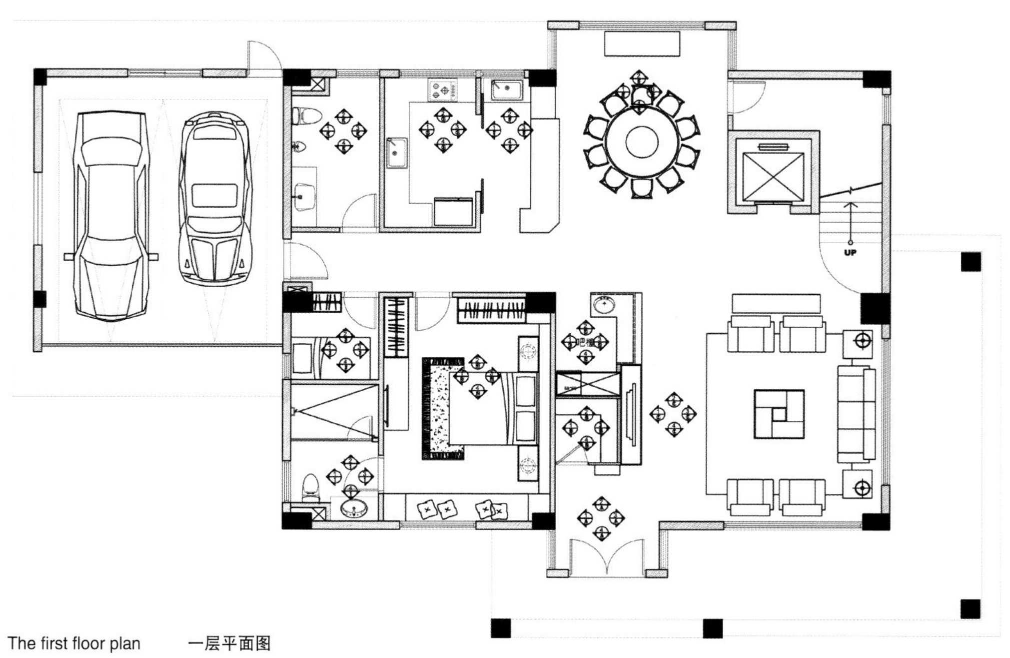

The first floor plan 一层平面图

此案充分考虑了窗户采光和坐落方向。房子有三个楼层，由私人电梯作连接。每一层都有其特有的机能性与私密性。一楼除了客餐厅为主要动线外，还有隐蔽的父母房和佣人房。客餐厅呈现新古典风格，运用石材做地面与墙面的造型切割。欧式风格的家具，营造低调奢华的气息，符合业主本身的大气风范。二楼则是起居间和家庭成员的房间。与一楼不同，二楼的设计是现代简约风格，虽然极简却不失典雅。看似黑白相间的壁面却是由不同材质来呈现黑色调性，赋予它低调却层次丰富的气质。三楼则是以家庭剧院、茶道室为主，让家人得以凝聚在此共度欢乐时光，并且有各自的娱乐场所。茶道区分为开放区及和室区，除了泡茶亦可当作客房使用。与一般住房不同的是，因为住家光线明亮、景色优美难得，设计师巧妙使用白银杉来作为和室的主体色系，让整个空间充满清幽高雅的气息。

ROME
SWISS
SPAIN
CZECH

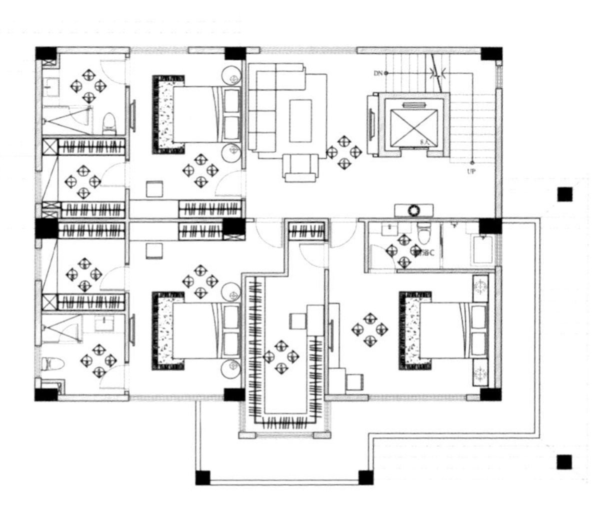

Second floor plan 二层平面图

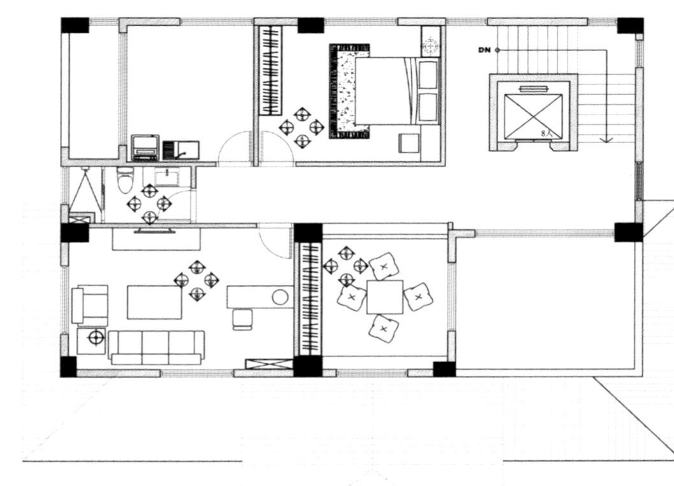

Third floor plan 三层平面图

Elegance & Tranquility

静宅·雅舍

- Design Agency: JuChen Design
- Designer: Ye Qiang
- Location: Fuzhou, Fujian
- Area: 500m²
- Photographer: Zhou Yuedong

- 设计单位：巨辰设计
- 设计师：叶强
- 项目地点：福建福州
- 项目面积：500m²
- 摄影：周跃东

It's a typical design of a male designer that is direct and practical. Under the simple façade with masculine features, every detail is showing the individuality and temperament. On stepping into the public area, you are welcomed by the living and dining room. The modern style manifests in the expression of the spacial temperament, by well combining the lights and colors. The brown palette seems to dissolve in the light, reducing the tedium and creating a tranquil modern atmosphere. Clean lines are used in the whole design to show the client's personality and deliberation, adding fashionable feelings to the space. The calm palette, the direct and natural furniture, a large area of direct lines used and concise and fluent space layouts highlight the whole space. The main bedroom can function as a bedroom, play house and office simultaneously, which shows the designer's reason and wise. The concise space layout enables the lights to flood into the room by eliminating visual obstructions, and create a relaxing and tranquil living space, like the ocean, like the galaxy.

这是一个属于男性设计师特有的设计，它透明直接、舒适实用。淳朴阳刚的外表下蕴藏着精细的心思，每一个细节都是有个性、有态度的。进入公共区域，首先映入眼帘的客厅和餐厅。设计师的现代主义风格主要体现在空间气质的呈现上，线条简约的元素通过色彩和光线有机结合，灯光下晕开的褐色像散开的水墨，消解了单调感，营造出独具现代感的沉静氛围。裸露于表面的线条组合苍劲有力地穿插在整座别墅的设计之中，发布着个性宣言，表明了从容不迫的态度，给这一空间增加了有别于传统的时髦味道。

冷静的色调、直接的摆设、大面积的直线条和简约通透的空间布局是这个家最显著的特点。主卧室的陈列将作息、休闲和办公三种功能空间一并囊括，一气呵成的设计非要有足够的理性功底才能驾驭。凝练的空间布局消除了视觉障碍，让光线充盈在开放式的空间内，构成了静谧松弛的主题，如深海，似星河。

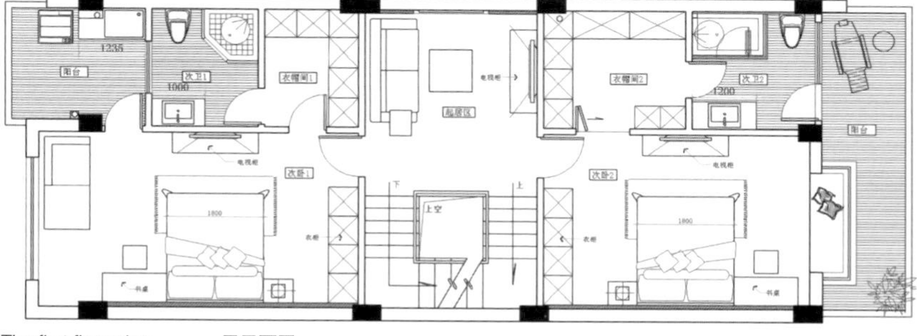

The first floor plan　　一层平面图

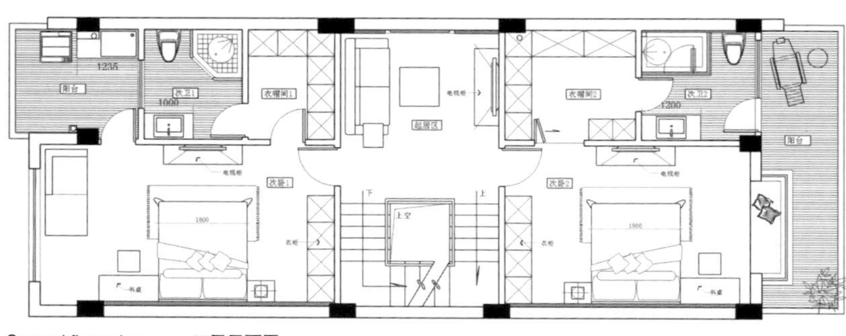

Second floor plan　　二层平面图

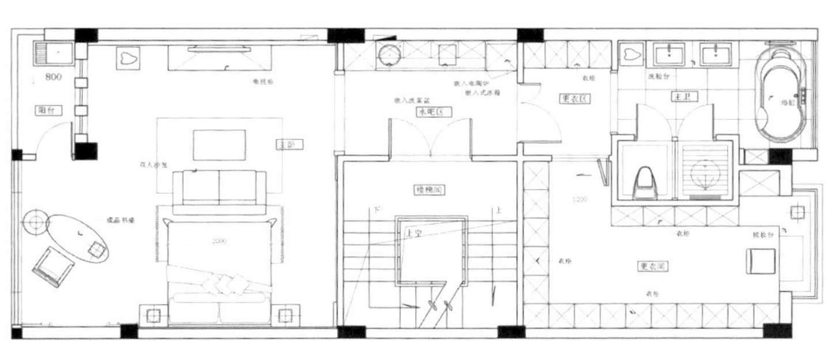

Third floor plan 三层平面图

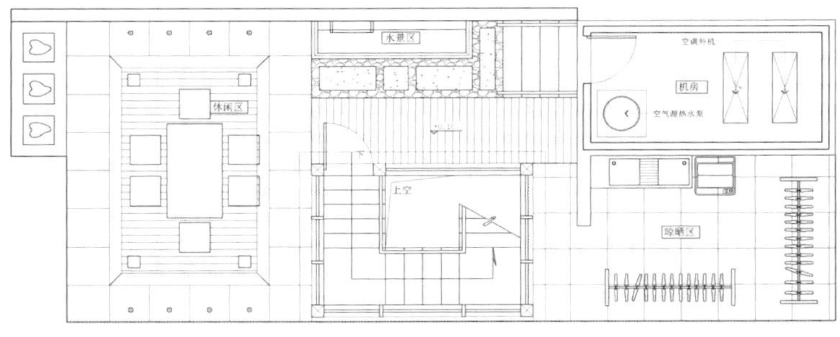

The Terrace of plan 露台平面图

Cliff House

悬崖之墅

- Design Agency: SAOTA, ARRCC
- Location: Dakar, Senegal
- Area: 1954 m^2
- Photographer: SAOTA

- 设计单位：SAOTA 建筑事务所，ARRCC 室内设计机构
- 项目地点：塞内加尔，达喀尔
- 项目面积：1954m^2
- 摄影：SAOTA 建筑事务所

Built on the site of an old World War Two bunker and on the edge of a cliff, Cliff House maximises its commanding position to create a house that is not only dramatic but with the incorporation of historical elements quite magical and mysterious. Part of the old bunker has been retained and a portion of it now houses an underground cinema that opens up into a water courtyard /moat that runs along the boundary creating a water feature at the gateway to the property. It is connected back to the house via a timber panelled walkway leading to a spiral staircase that runs from the lower ground through to the first floor and second floor levels of the villa.

The ground floor of the house, designed to facilitate seamless indoor and outdoor living and entertainment, is arranged in an L shape around the pool, the pool terrace and the garden. The formal Living and Dining spaces cantilever over the cliff and hang over the Atlantic Ocean enjoying panoramic sea views as well as views back to the house. The Kitchen made up of a so called 'American' or open kitchen and a separate traditional kitchen as well as the garage and staff facilities run along the east west axis and along the northern side of the boundary.

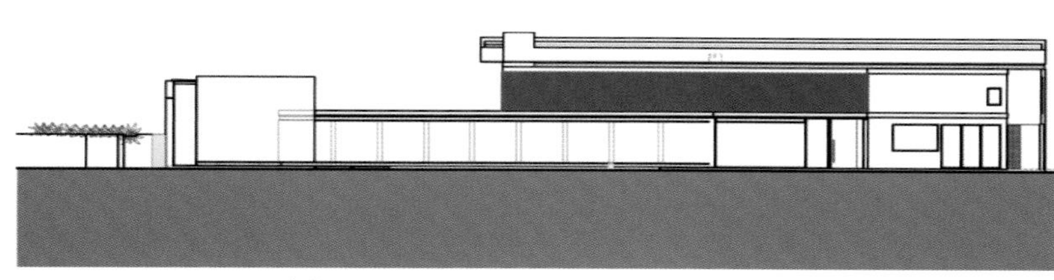

North East Elevation 东北面立面图

LEGEND

1. CINEMA
2. SERVICE AREA
3. BEDROOMS
4. PONO
5. GYM
6. KITCHEN
7. ENTRANCE
8. LOUNGE
9. STUDY
10. POOL
11. TERRACE
12. GARAGE
13. ENTRANCE GATE
14. GATE HUOSE

South East Elevation 东北面立面图

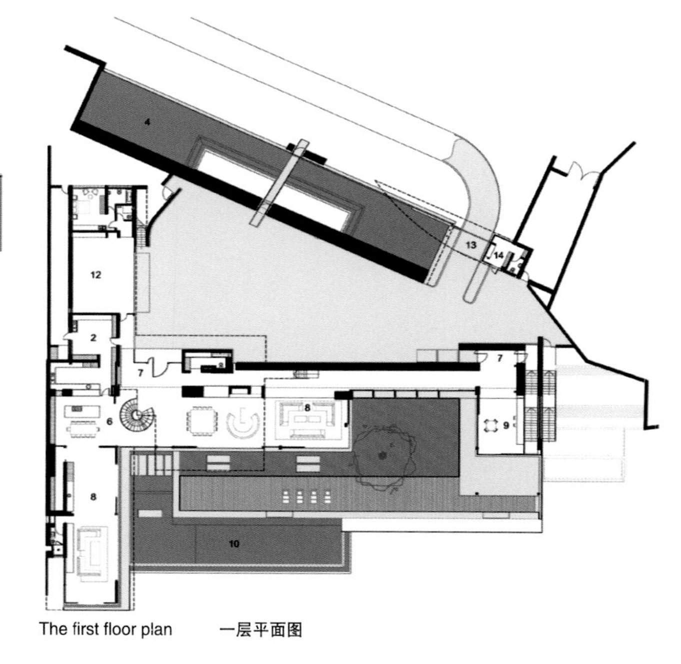

The first floor plan 一层平面图

One of the features of the house is the spiral staircase, clad in stainless steel, while the treads are clad in white granite. To add to the sense of continuity between the levels the 20mm in diameter stainless steel rods run from the first floor handrail to the lower ground floor, thus making the stairwell look like a sculptural steel cylinder. A skylight above the stairwell as well as floor to ceiling glazing in the lounges adds to the sense of transparency.

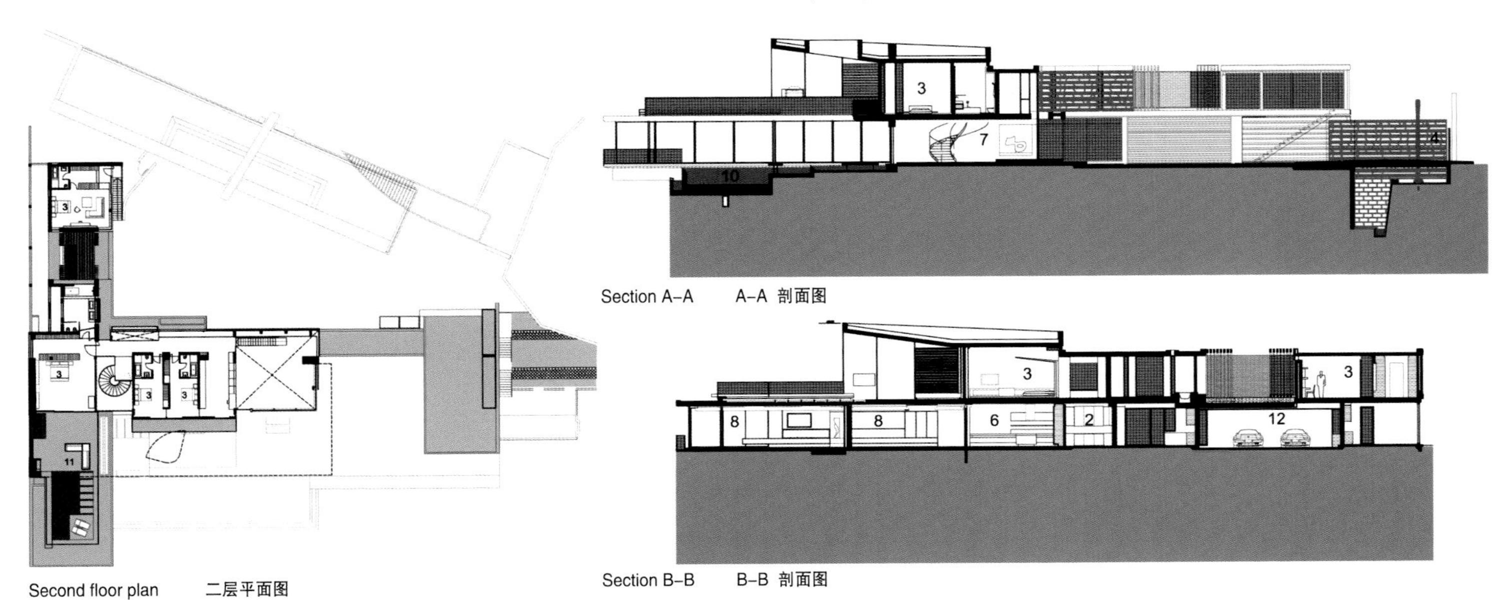

Second floor plan 二层平面图

Section A–A A–A 剖面图

Section B–B B–B 剖面图

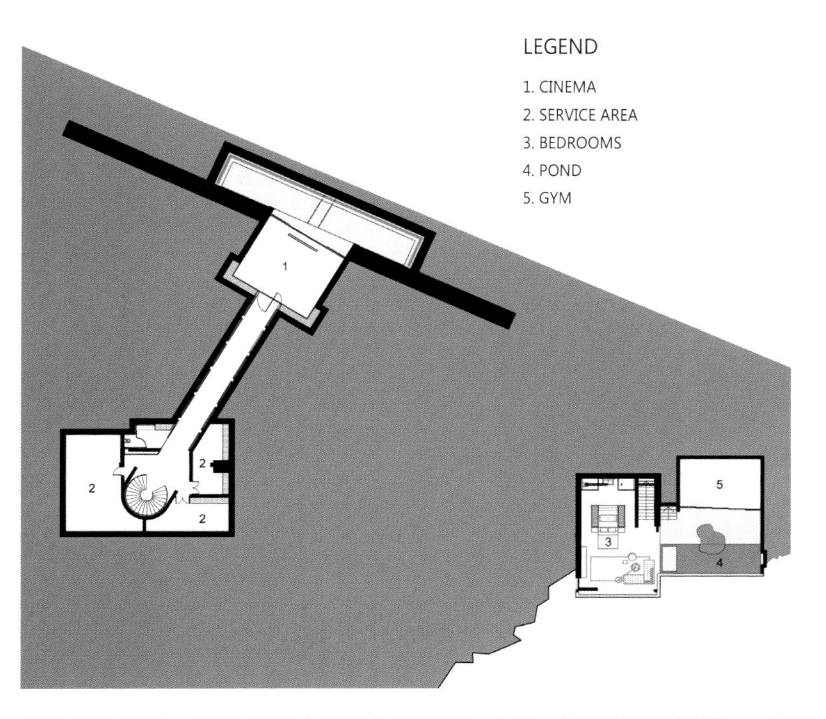

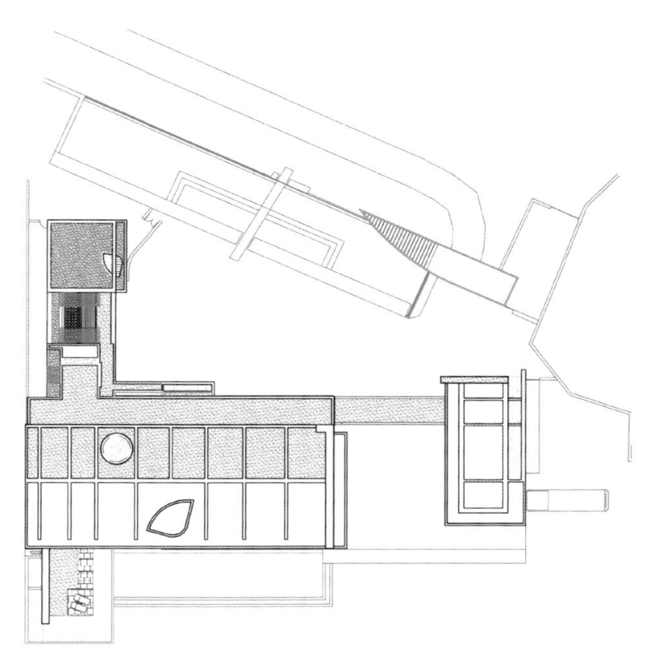

本案是位于悬崖边的一座别墅，其所在之处曾是“二战”时期的一个地堡。本案充分发挥地形优势，建造了一座完美结合历史特色和生动元素的别墅。

地堡的部分空间得到保留，其中一部分如今已被用作地下影院，面向一条战时的护城河，河水缓缓穿过地界，成为装点大门的一处水景。一条木板铺就的过道通向一座螺旋向上的楼梯。一楼空间围绕泳池和院子呈 L 形展开，形成了室内外休闲活动空间的自然过渡。客厅和餐厅似是悬在崖边海上，远可眺大西洋的美丽海景，近能赏悬崖之墅的别致风光。沿地界北侧绕东西轴的空间分布有员工设施和车库，以及一个美式的开放式厨房和一个传统风格的厨房。

螺旋楼梯是房子的一大亮点，拥有不锈钢包覆的扶手和白色大理石的台阶。为了增加楼层间的连贯美感，楼梯扶手采用了直径 20mm 的不锈钢细棒连接楼层，使整个楼梯井看起来像是一个别致的钢制筒状雕塑。而楼梯上方的天窗和落地窗则增加了空间的透光度。

Clovelly House

Clovelly 别墅

- Design Agency: Rolf Ockert Design
- Location: Sydney, Australia
- Area: 490m^2
- Photographer: Sharrin Reece

- 设计单位：Rolf Ockert 设计工作室
- 项目地点：澳大利亚，悉尼
- 项目面积：490m^2
- 摄影：谢林・里斯

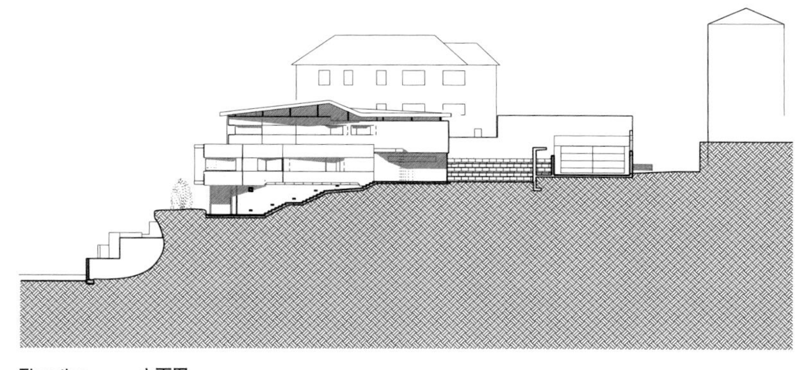

Elevation 立面图

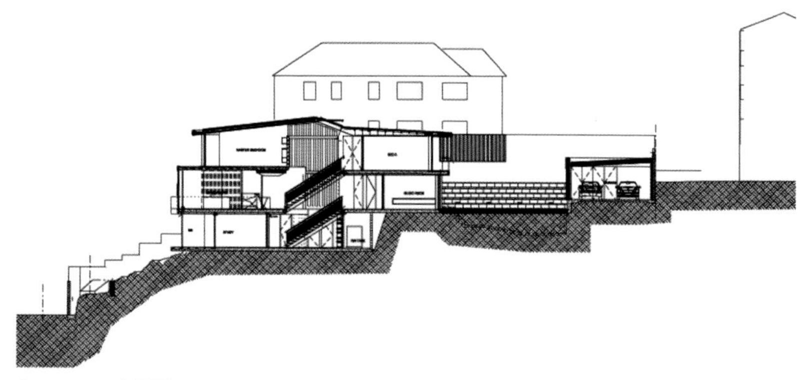

Section 剖面图

The unusual but elegant roof shape allows sunlight in while still allowing neighbours to enjoy water views over the lower end. The expressive angled concrete wall mirrors the roof shape but in negative, resulting in a complex facade geometry along the main face, enhanced by the movement of ever changing shadows over the shapes.

The light void also contains the central circulation, the stairs. These are light and airy without looking or feeling flimsy. To the North of the area two levels, to the South three, taking advantage of the natural slope of the site. The main living space is on the Entry level, connecting it with the Northern garden and Pool as a very generous central family area. Upstairs are the bedrooms, on the Southern lower level several areas for more individual activities, Study, Studios and Library.

As a consequence of the relentless Southerly winds the house was designed, unusually and against our original instinct, without any opening windows facing South. Instead large frameless floor-to-ceiling double glazed elements allow uninterrupted views over the Pacific and allow a more intimate visual connection than framed openable glazing elements would have afforded.

An outdoor deck is attached to the side of the Living area, allowing outdoor activity on suitable days without interruption of the front row feel the house enjoys.

The original Southern slope in front of the house was full of building rubble from some previous building incarnation. Once that was all removed several large natural sandstone blocks that had fallen eons ago and stood upright, affording us an unexpected, giant Japanese-style rock garden.

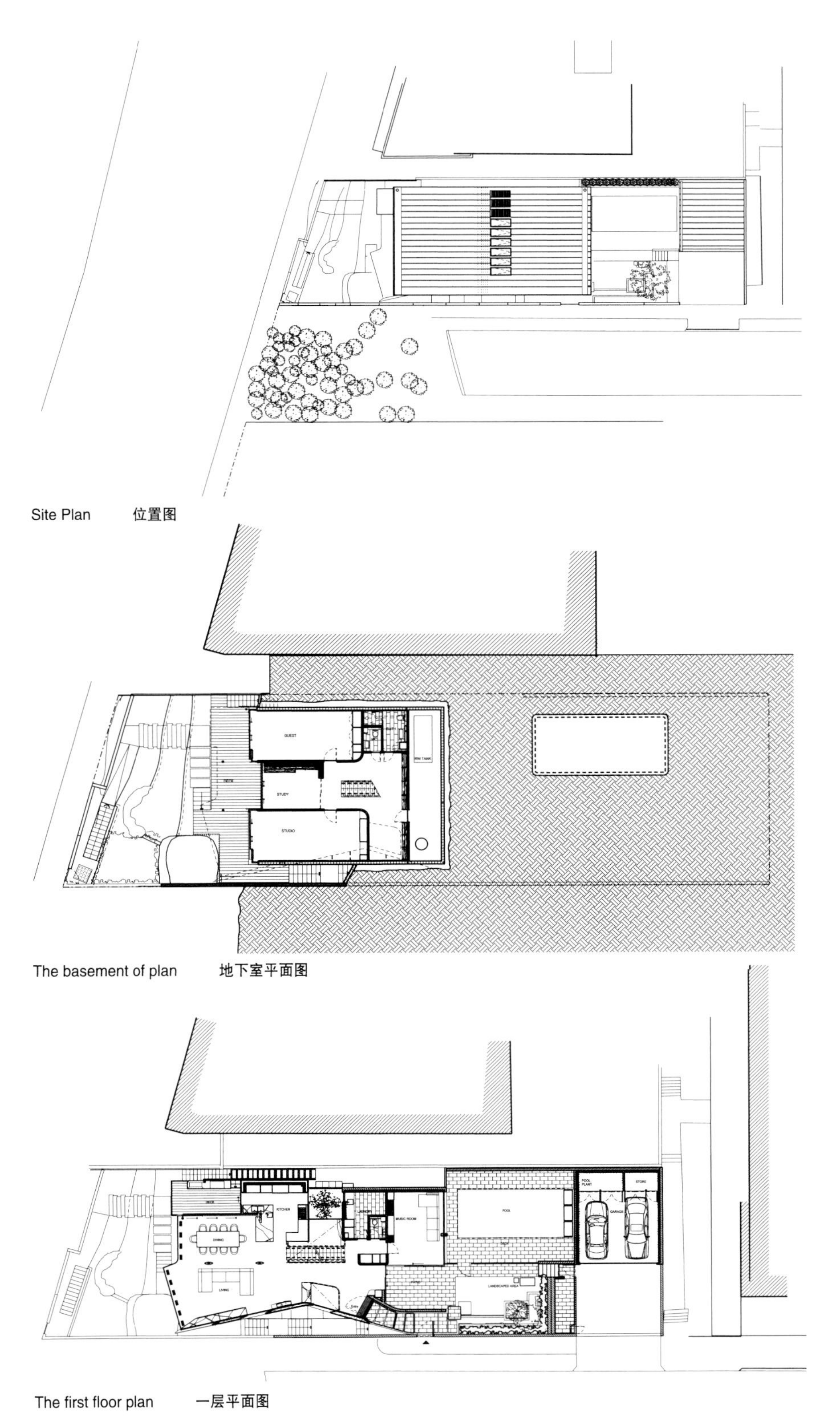

Site Plan 位置图

The basement of plan 地下室平面图

The first floor plan 一层平面图

本案是一栋海边别墅。造型优雅而别致的屋顶使得阳光得以进入室内，且不影响边上建的较低的房子的采光采景。阳光将屋顶的投影映在角度特别的水泥墙上，随着阳光和角度的变换，在上面映上了一幅幅丰富有趣的几何画面。设计师还充分利用建筑所在地的自然坡度，因地制宜地在北边建了两层楼，在南边建了三层楼，以此来改善房子的采光和通风条件。

别墅的主要活动区域设置在一层，和北边的花园以及泳池相连，为住户提供了非常宽敞自由的家庭活动区。卧房设在楼上，朝南的二层有书房、工作室和阅览室。因为南边有风终日不断，原本的房屋结构在朝南方向未开任何一扇窗，而这样的结构有违设计师的设计理念。因此，一扇无框的双层玻璃落地窗取而代之，使其既挡风雨，又能供人欣赏绝美的海景。屋外的甲板，可供人在适宜的天气走出室内充分享受户外的风景。房前南边的斜坡，堆放了之前建房时剩下的碎石，将之移走，铺上几块天然砂岩石板，将庭院打造成一个日式风情的岩石庭院。

FOR YOUR EYES ONLY

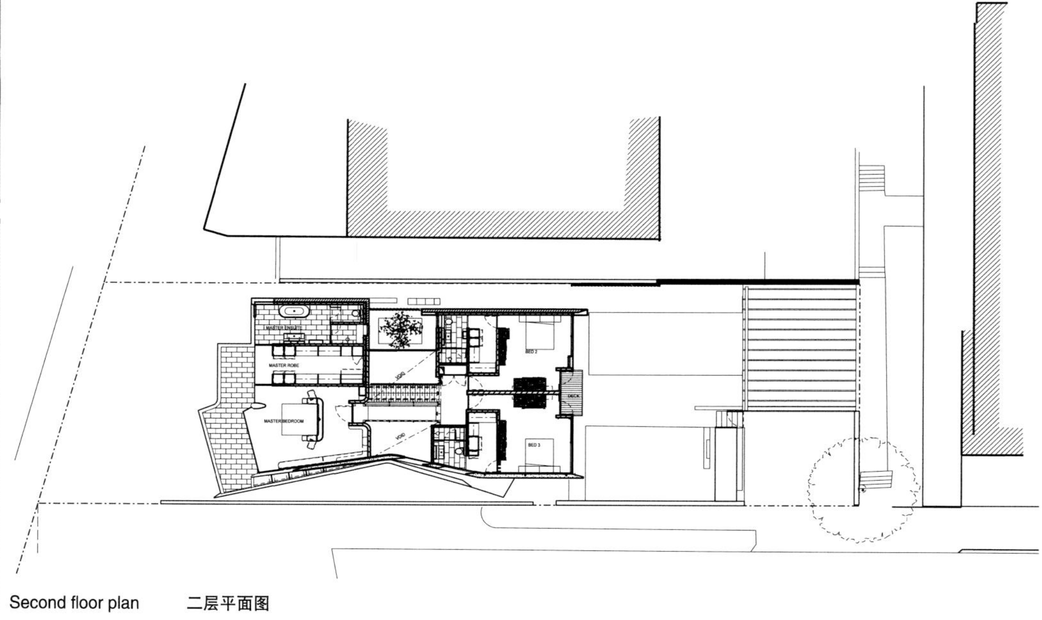

Second floor plan　　二层平面图

11 RMS

11RMS 别墅

- Design Agency: Pardini Hall Architecture
- Designer: Elisa Pardini, Robert Hall
- Location: London, England
- Area: 100m²

- 设计单位：Pardini Hall 建筑事务所
- 设计师：艾莉莎・帕蒂尼，罗伯特・霍尔
- 项目地点：英国，伦敦
- 项目面积：100m²

11RMS is a mews located in the heart of Knightsbridge village.

The internal planning responds to particular needs of the occupants, with the design concept driven by the desire to connect all three floors with one staircase. The stair is located centrally within the space and offers a different function at each level. At ground level its solid providing storage and division with concealed doors. Level one houses a bookshelf which together with glazed balustrading allows transparency, and at level two natural light to floods into the stair well and though the rest of the house.

The living spaces are designed as open spaces to allow natural light to enter through full height windows and the use sky lights. Light is a dominant theme both natural and artificially. Designed in collaboration with Viabizzuno, careful consideration to the warmth and color between spaces was key to the success of the scheme. The ground floor is used as a studio, for this reason the staircase, fulcrum of the project, has sliding panels integrated within the structure to divide the space between office and living - when required. The facade is left unchallenged to preserve the visual integrity of the mews as a street.

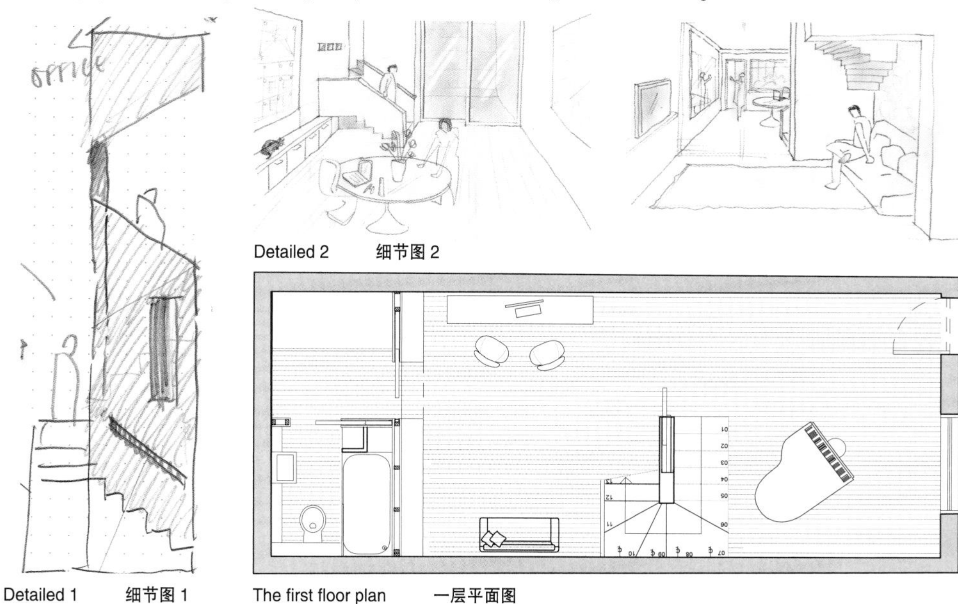

Detailed 2 细节图 2

Detailed 1 细节图 1

The first floor plan 一层平面图

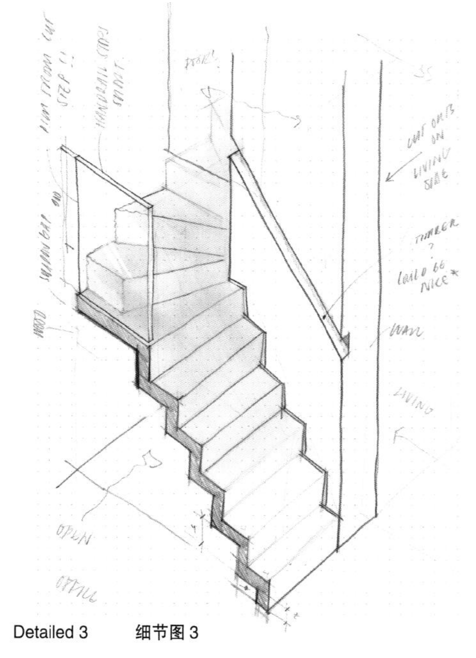

Detailed 3 细节图 3

该案是位于骑士桥村中心的一个由马厩改成的房子。

业主特别要求，要用一部楼梯连通三个楼层。设计师将位于房子中心的楼梯，在不同的楼层赋予不同的功能。设计师在一层楼梯旁的空间设置了一个书柜，将楼梯围栏换成全透明的玻璃；二层光线则透过楼梯的玻璃扶手涌入整个空间。

开放的空间设计使自然光线得以透过落地窗和天窗进入房间，而这些自然或人为的光线正是本案的主题。在与 Viabizzuno 的合作下，设计师对空间的温度和色调进行了全面考虑设计，这也是该案成功的关键因素。

位于最底层的工作室，在楼梯间安装了滑动门，以便在有需要的时候，将工作空间和居住空间分隔开来。房子的外观左侧并未做任何改动，意在保留原有的街头马厩的视觉效果。

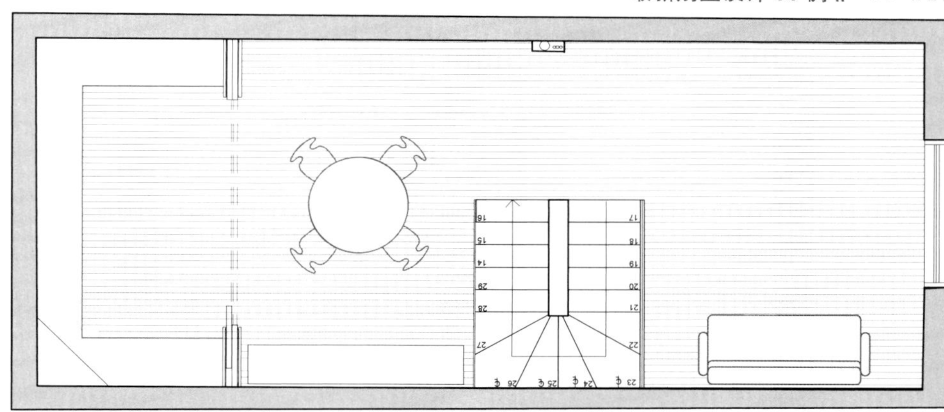

Second floor plan 二层平面图

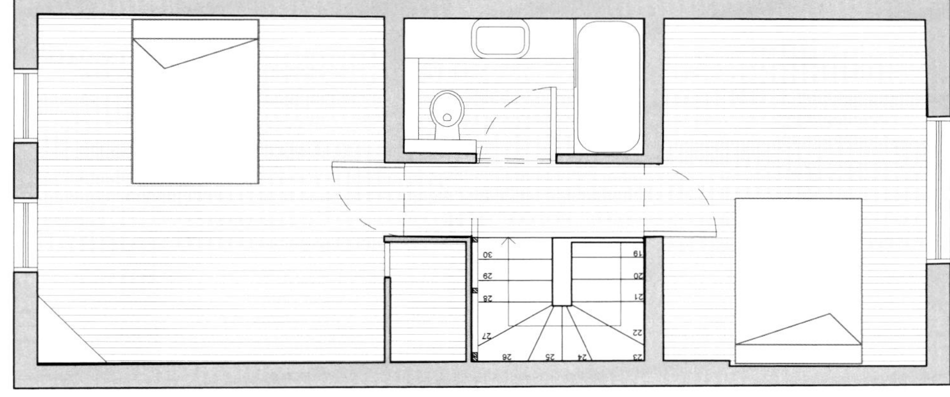

Third floor plan 三层平面图

Shenzhen Mangrove Bay Villa –Model House of Modern Style

深圳红树湾现代风格别墅样板房

- Design Agency: KSL Design
- Designer: Koon Shing Lam
- Location: Shenzhen, Guangdong
- Area: 775m^2

- 设计单位：KSL 设计事务所
- 设计师：林冠成
- 项目地点：广东深圳
- 项目面积：775m^2

Abandoning excess and complicated decorations, the design decorates the ceiling and walls with ivory-white oak finishing, echoing with marble's texture, to create a fresh and elegant atmosphere.

The raised space and clever layouts display the villa's character. Imported furniture of high quality adds more cultural qualities to the modern style space. The living room with the space high raised, is decorated with soft lights from imported lamps and high-end leather. The main palette of the study is log color, along with grey, to create a tranquil and cultural atmosphere. Fresh blue highlights the interior design that enhances the space. On the negative layer, the wine and cigar bar with an antique shelf and luxury leather furniture, cooperating with bright crystal lamps, shows the classical feels and the modern style.

此别墅样板房设计案例去除繁复的藻饰，雅白与橡木饰面的顶棚墙面搭配大理石的色泽纹饰，清雅的自然韵味内蕴深长。挑高的空间设计以及灵活巧妙的布局，形塑大宅的稳健气度。极富质感的进口陈设架构出空间的精致度与人文气质，焕发多重文化意蕴神采，晕染出深隽内敛的现代简约空间。

客厅中挑高的空间设计顿显明亮大气，在柔和灯光的调谐下，精美奢华的进口灯具与皮革等高档材质相互辉映。书房以木色为主、灰色为辅的基调营造了古朴安静的书香氛围，清新的蓝色给予视觉另一番惊喜，简洁而有力度的设计提升了空间的格调。地下一层的红酒雪茄吧，博古架和奢华皮革家具带来文化的多重韵味，璀璨剔透的水晶灯让古典韵味与现代简约巧妙兼融。

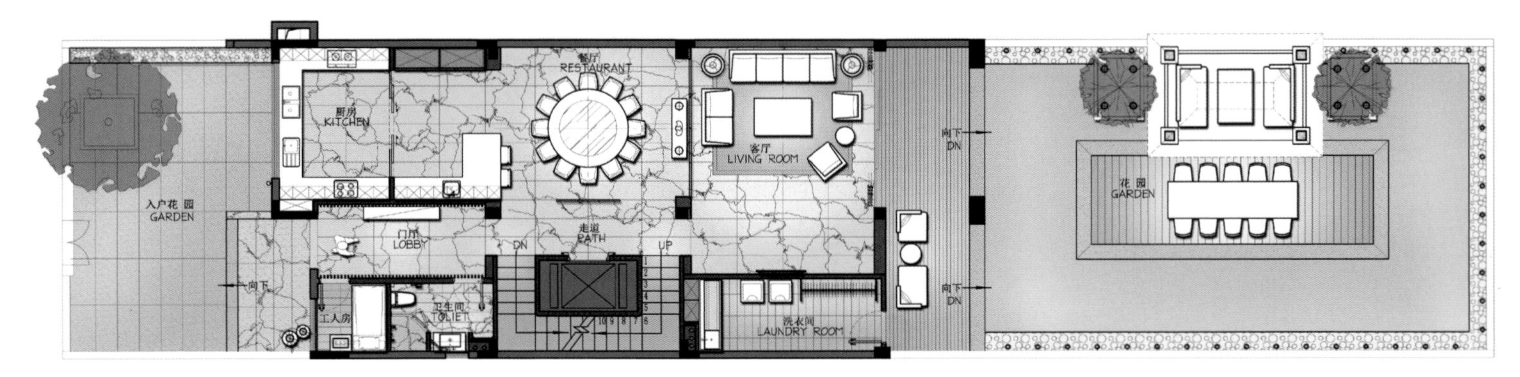

The first floor plan　一层平面图

Second floor plan　　二层平面图

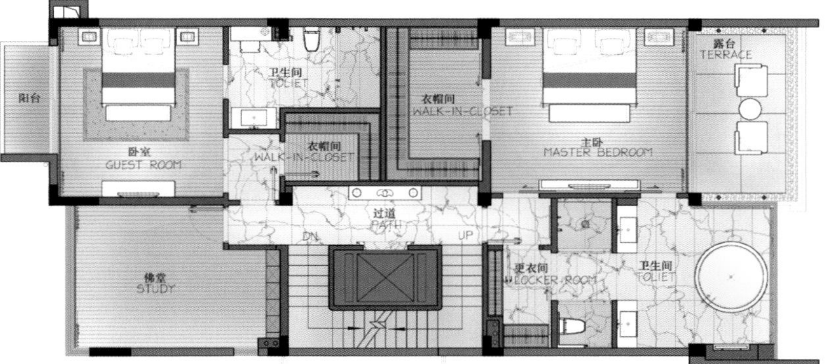

Third floor plan 三层平面图

The basement of plan 地下室平面图

Chalet 1.0

林中小屋1号

- Design Agency: YOD Design Lab
- Designer: Vladimir Nepiyvoda, Dmitriy Bonesko
- Location: Poltava, Ukraine
- Area: 140m^2
- Photographer: Andrey Avdeenko

- 设计单位：YOD 设计室
- 设计师：弗拉基米尔·涅皮沃达，德米特里·波尼斯克
- 项目地点：乌克兰，波尔塔瓦
- 项目面积：140m^2
- 摄影：安德烈·阿夫迪恩科

The apartments are the part of Relax Park Verholy, located in the middle of pine forest in Poltava region. The building consists of two rooms and a common guest lounge. Its architecture emphasizes the discreet and friendly nature around. The communication with the nature is underlined by the choice of natural materials, by the style of a mountain chalet and by restrained colors. The apartments come in a harmonious dialogue with the environment thanks to wood and stone used in a project. For decoration designers used basic elements of style "chalet" and realized them with a modern twist. Dominating warm shades of wood balance with the deep gray color applied for fireplace and in textile elements. This is a nice background for furniture, sanitary ware and decoration, made in a minimalist aesthetic.

该别墅位于波尔塔瓦的一片松树林中，是 Verholy 休闲公园酒店的一部分。别墅由两个房间和一个会客厅组成。重视与自然的交流与和谐共处，设计师特别采用天然材料，以沉稳的色调打造一座林中小屋。木石材料的应用实现了别墅与周边环境的和谐对话。在装饰上设计师遵循“木屋”的概念，使用了木屋风格的装饰元素，并且融入现代特色。原木的温暖色调平衡了壁炉和布艺沙发的冷灰，而配饰和家具等的自然组合充分体现了极简主义的美感。

Chalet 2.0

林中小屋 2 号

- Design Agency: YOD Design Lab
- Designer: Vladimir Nepiyvoda, Dmitriy Bonesko
- Location: Poltava, Ukraine
- Area: 120m^2
- Photographer: Andrey Avdeenko

- 设计单位：YOD 设计室
- 设计师：弗拉基米尔·涅皮沃达，德米特里·波尼斯克
- 项目地点：乌克兰，波尔塔瓦
- 项目面积：120m^2
- 摄影：安德烈·阿夫迪恩科

The guest house is designed in Alpine chalet esthetics and is a part of spa complex Relax Park Verholy. The designers have created a place where guests can relax in a pine forest away from the city. For more unity with the nature designers used natural materials and neutral colors in the interior and in the exterior. The house has two bedrooms with a common guest hall with panoramic windows and a fireplace that can be used from the house and from the terrace. The interior furnishing is dominated by natural stone and wood. Its linear laying creates a visual rhythm, combined with pine trees outside. Near the house there is a swimming pool and sitting areas.

作为 Verholy 休闲公园酒店的客房，这栋别墅被设计师打造成了一座别致的山中小屋。位于松林之中的别墅，为前来的每一个客人提供了一个远离城市喧嚣的休闲之所。为了使设计更贴近自然，设计师在室内外设计中采用了自然材料和温和中性的色调。别墅由两个房间和一个会客厅组成。会客厅拥有一扇全景落地窗和一个内外两用的壁炉。家具和饰品主要由木石材料制成，具有视觉韵律的线条，呼应着屋外的松树。屋外还有一个泳池和休息区。

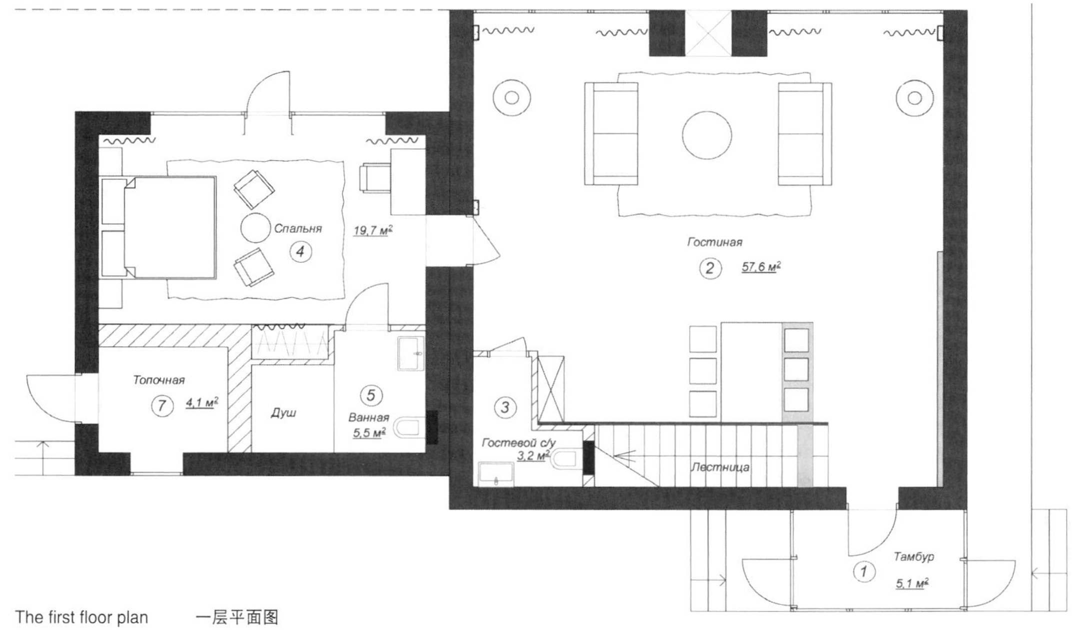

The first floor plan　　一层平面图

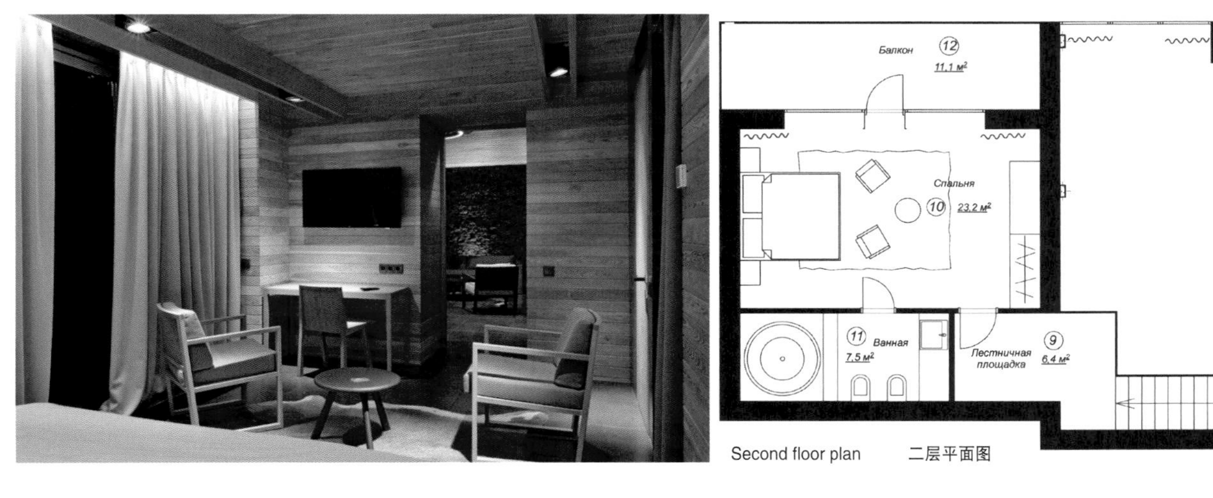

Second floor plan 二层平面图

Chalet 3.0

林中小屋3号

- Design Agency: YOD Design Lab
- Designer: Vladimir Nepiyvoda, Dmitriy Bonesko
- Location: Poltava, Ukraine
- Area: $140m^2$
- Photographer: Andrey Avdeenko

- 设计单位：YOD 设计室
- 设计师：弗拉基米尔·涅皮沃达，德米特里·波尼斯克
- 项目地点：乌克兰，波尔塔瓦
- 项目面积：$140m^2$
- 摄影：安德烈·阿夫迪恩科

The guest house is designed in Alpine chalet esthetics and is a part of spa complex Relax Park Verholy. The apartment consists of three rooms and a common living room with a fireplace and a large dining table. To create relax atmosphere the designers chose quiet, neutral colors and natural materials with a nice texture – skins, travertine, stone and wood. The house has panoramic windows. Part of the wall has a ribbed cladding made of plywood with backlight, and one of the walls is decorated by leather elements cut into squares.

这座小屋，同样是 Verholy 休闲公园酒店的客房，经过设计师的巧手而颇具山林木屋的美感。别墅拥有三个房间和一个会客厅。会客厅中有一个显眼的壁炉和一张大餐桌。为了营造轻松舒适的氛围，设计师选用了温和中性的色调和质感优良的用材，如皮革、石灰华、石头和木材等。会客厅还有一扇全景落地窗，让人即使在屋内也能尽情观赏窗外的美景。部分墙面由胶合板构成，其中一面墙装饰了许多方块皮革。

The first floor plan　　一层平面图

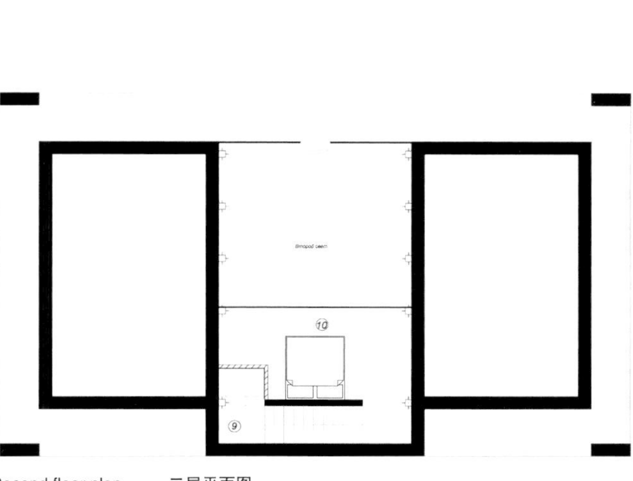

Second floor plan 二层平面图

Elena Dobrovolskaya House

Elena Dobrovolskaya 住宅

- Designer: Elena Dobrovolskaya
- Location: Dnepropetrovsk, Ukraine
- Area: 860m^2
- Photographer: Igor Karpenko

- 设计师：艾琳娜·多布罗沃斯卡亚
- 项目地点：乌克兰
- 项目面积：860m^2
- 摄影：伊戈尔·卡朋克

The interior style is based on elegant Art Deco luxury, artistic sense of it, and the peculiar mood of the splendid epoch. The house has four floors. The living room and the dining room are situated on the ground floor, the guestroom and the daughter's bedroom are on the first floor, and the second floor is feminine with its own infrastructure: bedroom, cloakroom, small cabinet and bathroom. And the third floor is a masculine territory with spacious cabinet with the collection of Netsuke miniature sculptures, bedroom, and subsidiary rooms. The female spaces are light and elegant while the man's rooms are more dark and restrained.

Ornaments set the rhythm and the mood for the whole interior, complying with the Art Deco tradition – geometric and vegetative decorations, applied on the walls and on the floor, on some ceilings, and in decorative details. The interiors are realized in placid monochrome scales of dark brown, terracotta, cream, sandy, and silvery grey tones. The visual accents

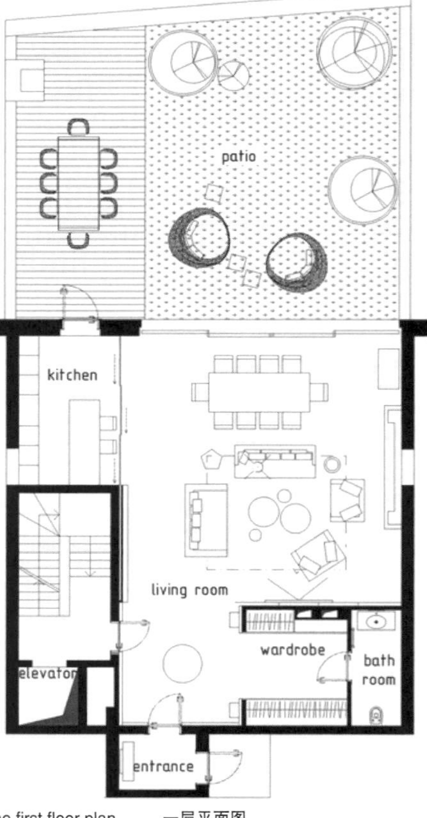

The first floor plan　一层平面图

are expressed with eye-catching pieces of furniture, lighting objects and various rich textures. Almost all decorative details are based upon individual sketches. The distinctive feature of this elegant interior is the complicated handicraft done by sculptors and wood, metal and stone engravers. Small Tuscan manufactures and Ukrainian workshops produced copper engraved boiseries, brass foldingscreens, stone-carved and plaster cast details, designed specially for this project.

Second floor plan 二层平面图

本案的室内设计具有 Art Deco 风格的优雅奢华，且注重艺术感和对那个辉煌时代的特殊情怀。

本案是一栋 4 层楼房。客厅和餐厅在一楼，二楼是客房和业主女儿的房间，三楼是更有女性特色的空间，包括卧房、衣帽间、小隔间和浴室。四楼则更具阳刚味：宽敞的放了 Netsuke 塑像的柜子，充满男性特点的卧房，以及几个隔间。三楼的设计优雅明亮，四楼的设计则更沉稳素净。室内设计所用的装饰品遵循 Art Deco 的传统，如几何图形和植物装饰的应用，见于墙壁或地板，以及一些顶棚等细节处，带动了整个设计的节奏和氛围。本案设计的色调偏于温和的单色，采用了棕色、赤土色、奶油色、淡茶色和银灰色。引人注目的家具、灯具和纹理丰富的各种装饰，共同呈现了一种特别的视觉效果。

设计师几乎为所有的装饰细节都描绘了设计草图。这个优雅的室内设计，仿佛是由雕塑家和雕刻家亲手完成的一件由木石金属工艺组成的艺术品。来自托斯卡纳和乌克兰工坊的铜制雕花镶板，黄铜制的可折叠屏风，用上了石刻和石膏模型的设计细节，让整个设计大放光彩。

Third floor plan 三层平面图

Fourth floor plan　　四层平面图

Villa Sofia

Sofia 别墅

- Designer: Elena Dobrovolskaya
- Location: Yalta, Crimea
- Area: 1150m^2
- Photographer: Igor Karpenko

- 设计师：艾琳娜·多布罗沃斯卡亚
- 项目地点：克里米亚，雅尔塔
- 项目面积：1150m^2
- 摄影：伊戈尔·卡朋克

A harmonious relationship of the exterior and the interior creates the wholesome character of the hotel. In Villa Sofia you can feel the atmosphere of unique combination of different styles and times. Each of the rooms has its own charisma, which is emphasized by elaborate details, colors and accessories. This is typical for all the objects of Elena Dobrovolskaya. You can experience the refined classic French charm or feel the magic breath of the East. While the combination of East and West is a favorite source of inspiration for many architects, Villa Sofia combines the best of both worlds and it is not a cliché.

The business card of this hotel is the exclusive use of natural materials in the decoration: natural stone, wood, marble. Most of the furniture in the rooms was created specifically for Villa Sofia by Italian designers according to individual projects. All of the hotel's spaces have wooden window frames, and each of the windows looks out to the gorgeous Crimean landscape and fills your lungs with fresh sea air. Arched windows and refined classic furniture coexist with modern comfort and amenities – such as plasma displays, mini-bars and room safes. Each room has luxury curtains and linens and a separate bathroom with a unique design, equipped to the highest modern standards.

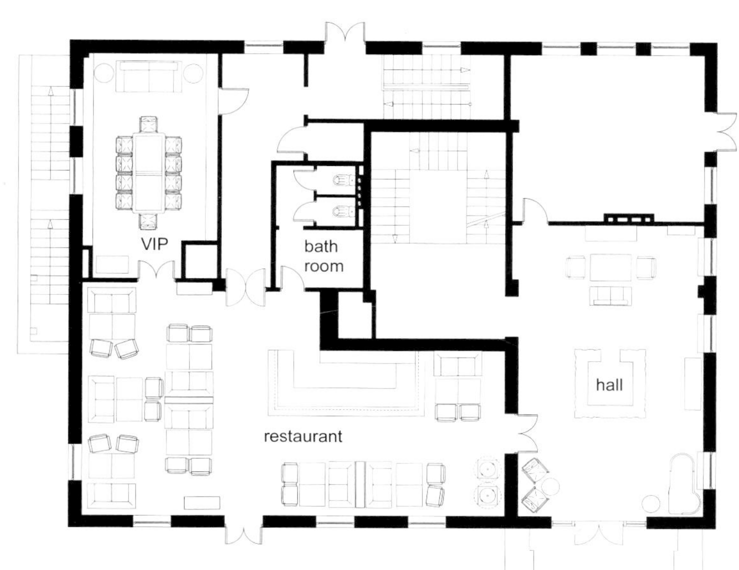

The first floor plan 一层平面图

The twelve guest rooms of the hotel are decorated in several refined styles. The classic French design and the Oriental, with Moorish style elements, are the most popular among the visitors. In the French style rooms you have a fresh palette, wallpapers with rich floral patterns and lightweight furniture, and in the Oriental interiors – blue purple maroon colors and lighting design typical of hot countries. The use of contrast colors in modern European interiors is softened by textiles and ornate decoration. Each room is different, yet, what remains unchanged is the use of natural materials and the care for environment.

本案是一间酒店。室内外设计的和谐搭配是本案的一大亮点。在这里能感受到一种独特氛围和不同时代及不同风格的融合。精致的设计细节、美丽的色彩和独特的配饰，无不彰显着每个房间各自的魅力。经典细致的法式风情，神秘丰富的东方魅力，都能在这里一一感受。东西方特色的融合是众多设计师的灵感之源，而 Sofia 别墅的设计将此特色发挥得淋漓尽致，独树一帜。

Second floor plan 二层平面图

Third floor plan 三层平面图

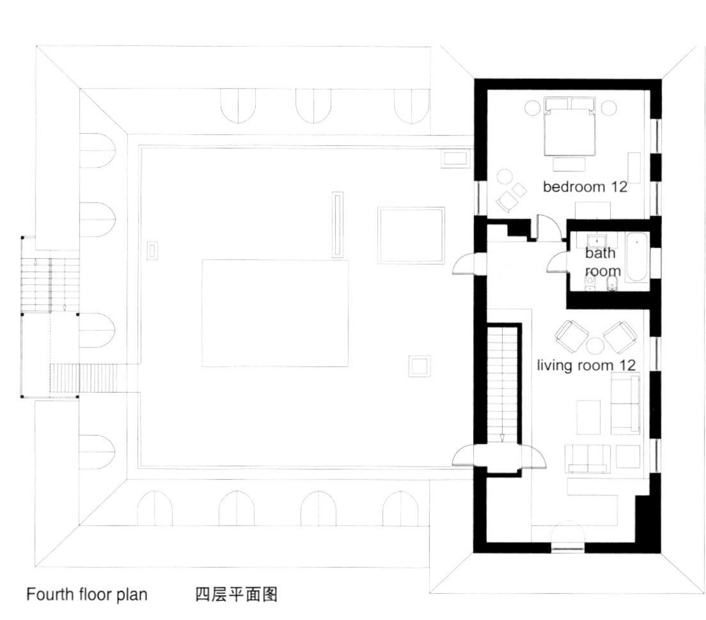

Fourth floor plan　　四层平面图

酒店的用材为独有的自然材料，包括石块、木材和大理石。房间中的大部分家具都是由酒店根据不同的风格向意大利设计师专门定制。客人可以透过每一扇木窗户欣赏克里米亚的美景，感受吹来的海风。还有拱形窗户和古典家具搭配最高标准的现代化设施如等离子电视、迷你吧台和保险箱，以及华丽的窗帘、床单和特别设计的独立浴室。

酒店的 12 间客房分别有 12 种风格的设计。其中，经典法式、魅力东方和摩尔文化最受客人的欢迎。法式风格的房间，色调清新，装饰有花纹丰富的墙纸和轻便的家具；东方韵味的房间，色调则偏向蓝色、紫色和栗色，一些具有热带国家特点的灯饰也颇具特色。现代欧式风格的房间设计使用了一些布艺和饰品来使房间色彩的对比趋于柔和，但不变的是注重对自然材料的应用和与跟周围环境的融合。

Cherry Garden

北辰小区樱花苑

- Design Agency: Zhongce Decoration
- Designer: Lin Lin
- Location: Kunming, Yunnan
- Area: 230m^2

- 设计单位：中策装饰
- 设计师：林琳
- 项目地点：云南昆明
- 项目面积：230m^2

The nature-respecting client who loves wood furniture and the leisure culture of tropical countries, has a happy family of three generations. Thus, the project's style is designed to be the south-east Asia style. The modern minimalism charm is integrated with the graceful and natural beauty, adding more colors and diversity, whilst remaining its features.

The concise lines add more calm feels to the space while the decorations display the south-east Asia style. The palette is medium yellow that makes the space warm and natural. The kitchen on the first floor consists of a Chinese kitchen and a Western kitchen, which not only fulfill different cooking requirements, but also enhance the dining atmosphere. The public area is divided into two parts-the living room and the leisure area. The balcony is dismantled and has been changed into the terrace and the garden, which broadens the view of the living room and borrows the scenery to make the space more lively, cultural and comfortable. To fully use the space, the stair on the second floor is improved and provided with more clear lines.

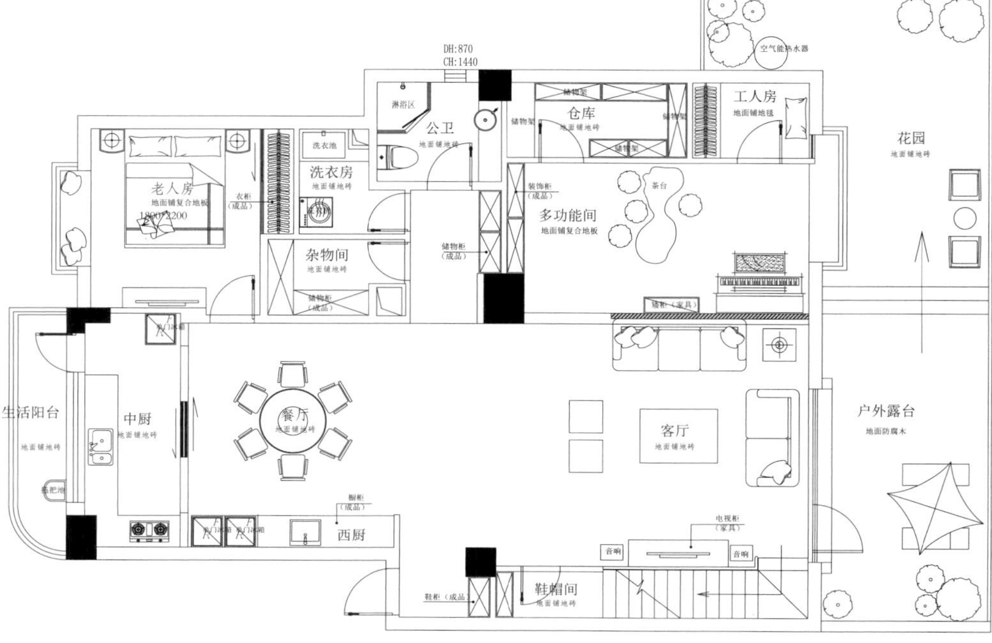

The first floor plan　一层平面图

本案的业主是崇尚自然、喜欢原木家居、喜爱热带休闲文化的成功人士，有一个三代同堂的幸福家庭。通过和客户沟通，我们将风格定位为东南亚风格。本案中东南亚风格在婉约、自然原始之美的基础上，还融入了现代简约手法，在保留自身特色之余，产生了更丰富多彩的变化。凝练的直线条，使空间看上去沉稳干练；细节处点缀充满东南亚风格的装饰，凸显其特征。将亚洲文化凝结成象征意味浓郁的系列符号。方案中主要色调为中黄色，整个空间看上去温馨、自然。在一层设计了中厨和西厨，这样不仅能充分保证业主对厨房的不同需求，还可以增加餐厅用餐氛围，使整个就餐环境变得惬意、温馨。一层公共空间把原本空旷的空间分成了客厅和休闲区，使得两个空间不仅具有围合感还相互呼应。将原本的阳台拆除，改为花园和户外露台，改后整个空间效果瞬间改善，不仅拓宽了客厅视线，还将室外的自然景色借到客厅让整个空间看上去更生动、更自然、更舒适。二层原有的楼梯改了以后，整个空间都能得到充分利用且具有清晰的流线。

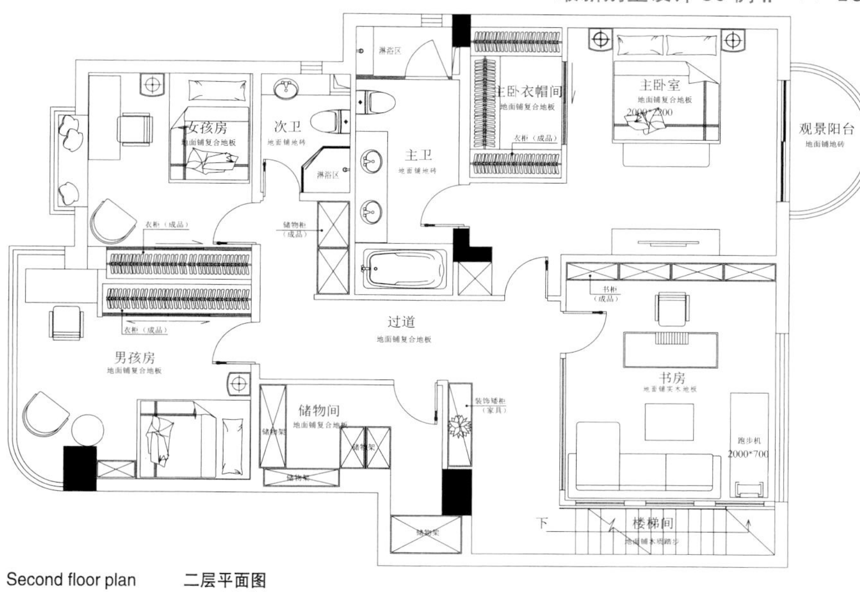

Second floor plan　　二层平面图

The first floor plan　一层平面图

Dianchi Xiangyang Garden

滇池向阳院

- Design Agency: Zhongce Deocoration
- Designer: Lin Lin
- Location: Kunming，Yunnan
- Area: 400m^2

- 设计单位 : 中策装饰
- 设计师 : 林琳
- 项目地点 : 云南昆明
- 项目面积 : 400m^2

The entire space is painted into yellow, white, golden and beige, those bright colors make this whole space pretty and bright. The neo-classic style of this house has a dual effect on both being fashionable and classic.

Public area in the ground floor has been changed: partition between original kitchen and dining room is removed and the kitchen is now open-ended. This can not only make sure there is enough space for cooking, but also create a relaxing atmosphere for family members having breakfast happily since early in the morning. Bedroom on the first floor is reconstructed as guest room and guest bath room, while the girl's room takes full use of space by opening a window in the study, thus making the room much brighter.

A partition is added to the original structure on the second floor; the pillar is hidden in the closet in the dressing room. The study in this floor is enlarged to meet functional needs of this space, as well as present the invaluable nobility and quality of the host.

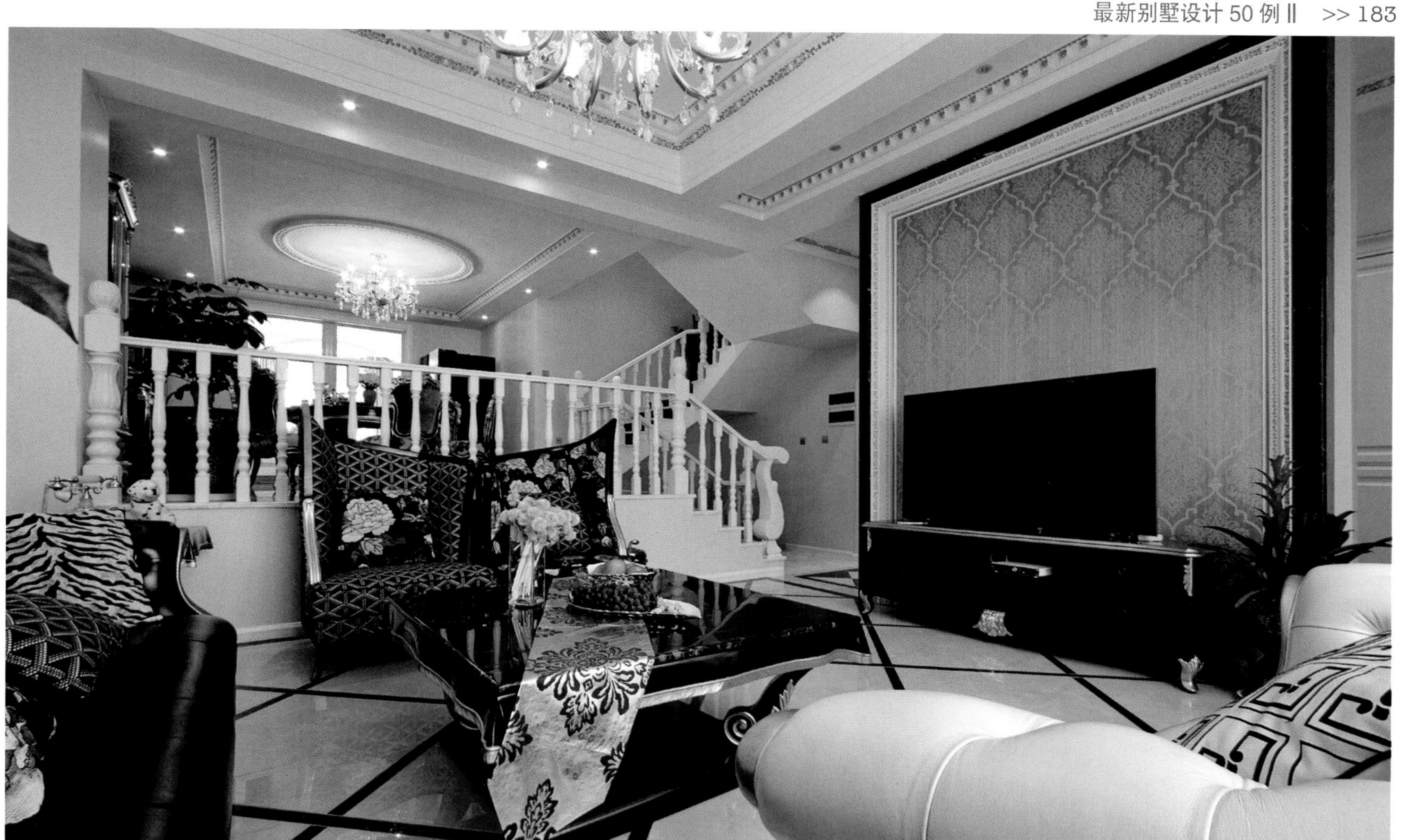

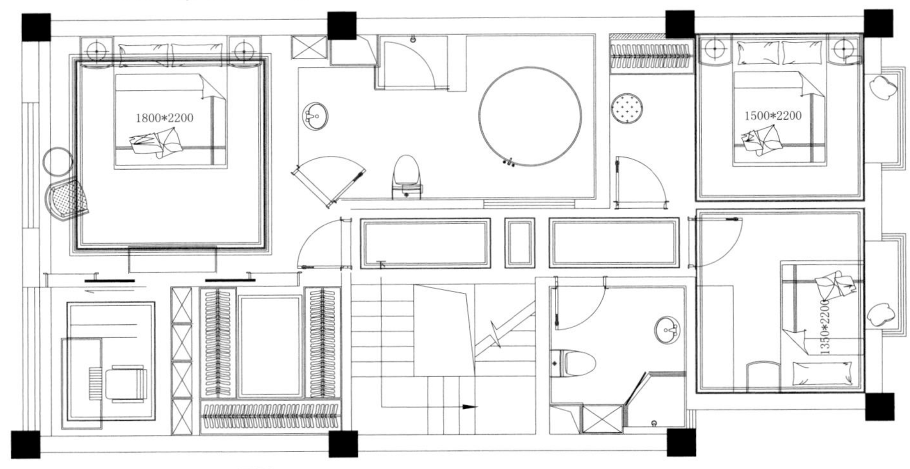

Second floor plan　　二层平面图

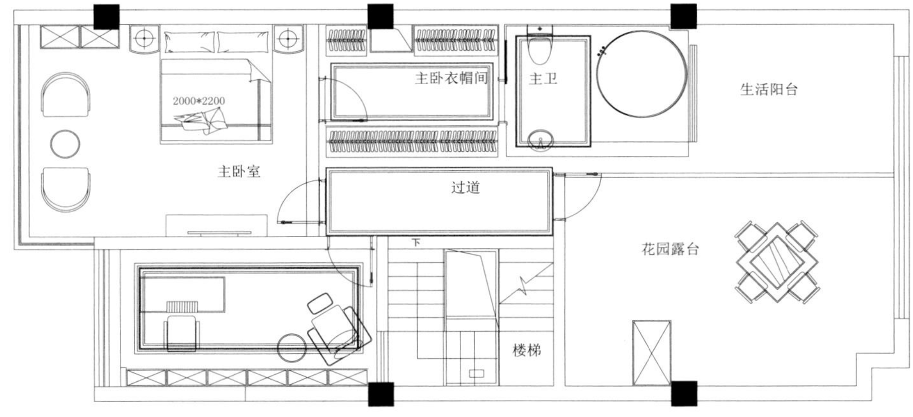

Third floor plan　　三层平面图

整个空间主要采用米黄色、白色、金色、黄色色调，大量揉入白色、明黄色与金色，使空间更呈现明快与靓丽。该案采用的新贵古典风格不仅具有强烈的时尚感还具备了古典与现代的双重审美效果。

一层公共空间，拆除原有的餐厅和厨房的隔墙，把厨房改为开放式厨房，这样不仅能充分保证厨房完整的操作台面，还可以增加餐厅用餐的小氛围，让居家生活的贴心快乐从清早开始就伴随全家人。二层把主卧隔为客房和老人房，将主卫改为公卫，实现了空间的巧妙利用。女孩房功能齐全，充分利用空间，书房开了窗户，让整个空间敞亮不少。三层主卧在原有结构上加了隔墙，还把原本主卧室的柱子隔到更衣间区域，通过衣柜对其进行隐藏处理。在原有结构上把三层书房扩大，改后整个空间在满足功能需求的同时也凸显出主人的尊贵与气量。

M House

M 住宅

- Design Agency: Marcel Luchian Studio
- Designer: Marcel Luchian
- Location: Singera, Moldova

- 设计单位：Marcel Luchian 设计工作室
- 设计师：马塞尔・卢西恩
- 项目地点：摩尔多瓦，基希纳乌

M House is a contemporary architecture based on overlapping rectangles designed for nice panorama view in Singera, Moldova.

So the shape is unique if we look for a second from the outside. With accents on facade forms, minimalist interior offer you correct function of it with light and shadow and materials. The house is situated with most living spaces enjoying maximum daylight via strategically placed ventilating windows to south, east and west-oriented window walls, providing a variety of exposures and views. The polished concrete floors on both floors have radiant heat pipes embedded in the structure, providing the primary source of heat. Energy-efficient heat pumps provide supplemental heat and air-conditioning on rare occasions. Spray-foam insulation was used on all perimeter walls, at the roof and between the first and second floors to maximize the thermal efficiency of the structure.

M 住宅是位于基希纳乌的一栋现代建筑，它拥有良好的视野，且造型独特，从外观上看，是由两个叠加的矩形空间组成。

其室内设计通过对材料和光影的巧妙运用，给人带来极简主义独有的美感。房子采光极佳，朝东朝南的窗户通风透气，西边覆盖了整个墙面的落地窗，让整个房子通明透亮。

抛光的混凝土地板下安装了嵌入式的热辐射管道，用于房屋的供暖。节能热泵则用于在有需要的时候提供热量补充和温度调节。墙面和顶棚采用了泡沫绝缘材料，用以提高结构的热效率。

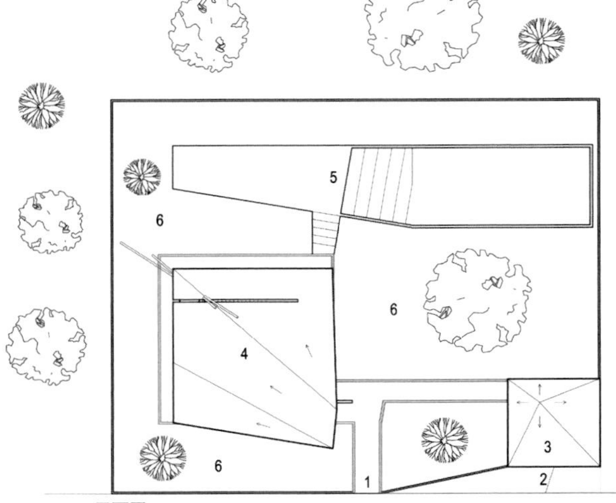

TERENUL

1 Intrarea pe teritoriu
2 Intrarea garaj
3 Garaj
4 Casa
5 Piscina
6 Spatiul verde

S terenului - 11.22 ari
S const. - 100 m²
S totala - 300 m²
S locuibila - 56.5 m²
V cladirii - 960 m³

S piscinei - 170.1 m²
S garajului - 47.5 m²

Plan 平面图

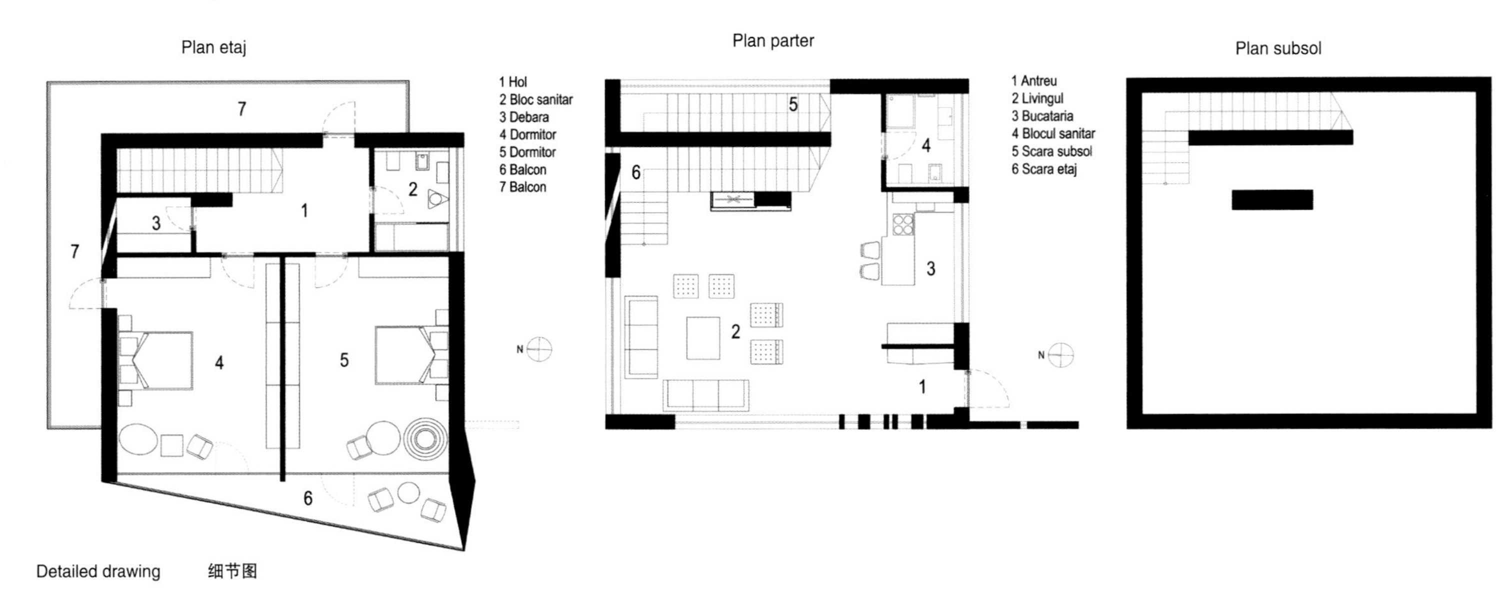

Detailed drawing 细节图

Villa Bayad

Bayad 别墅

- Design Agency: Popo Danes Architect
- Designer: Popo Danes
- Location: Bali, Indonesia
- Area: 7415m^2

- 设计单位：Popo Danes 建筑设计机构
- 设计师：波波・戴恩斯
- 项目地点：印度尼西亚，巴厘岛
- 项目面积：7415m^2

Located in Payangan, Bayad Ubud Bali Villa is in a rural location with local attractions. The strong point of Bayad Villa lies in the humbleness of the architecture's design. The villa bows down to nature by appearing here and there among the lush greenery of the landscape.

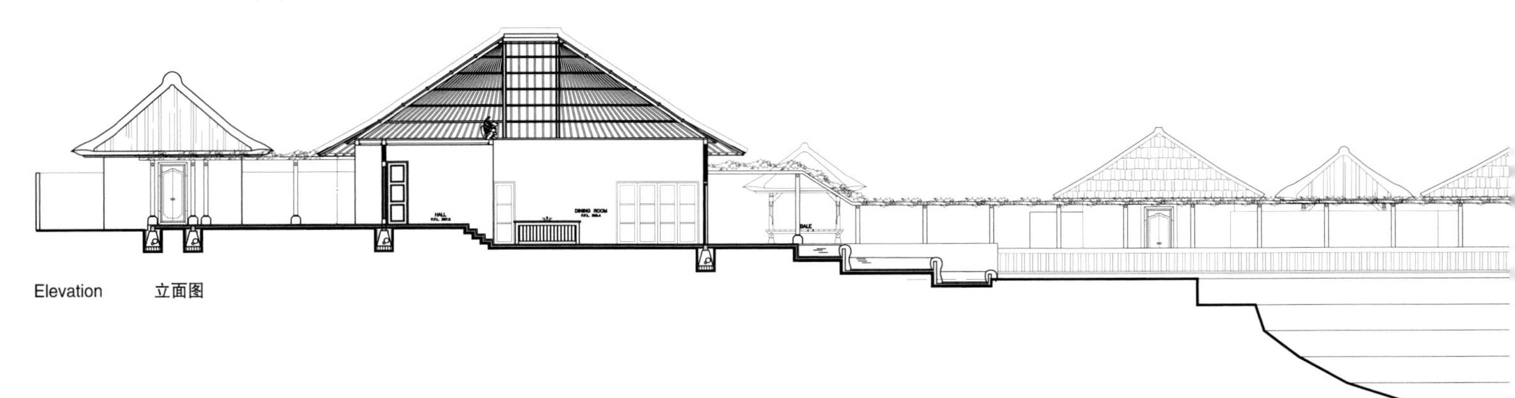

Elevation 立面图

On the almost 7500 sqm site, 6 buildings with 4 bedrooms are built in two opposite clusters. On one side are the two main cottages reserved for the owners of the resort, and on the other are two cottages to be rented. The rest of the buildings are used as common rooms. All buildings were built in such a way to intrude the natural surrounding as minimum as possible.

No massive buildings are built on this site. Most of the buildings are open buildings with slim wooden or concrete columns to support the roof. Gazebo and corridor connected one building with another. The whole design creates a cluster of manmade buildings that blends well and also creates an intimate "dialogue" with the surroundings.

本案别墅位于巴厘岛郊外的一个景区。设计师通过将别墅融入葱郁的绿色景观来表达其对自然的尊重。在这 7500m^2 的土地上，6 栋带有 4 间卧室的房子分列两旁。其中两栋是业主自留用作住房，还有两栋用于出租，剩下的则用作普通用房。每一栋房子的设计都尽可能少地改动周边的自然环境。开放式的建筑以木柱和水泥柱支撑屋顶，凉亭和走廊沟通各栋小楼。该案的整体设计使各栋小楼相互配合，实现了与周边自然环境的和谐对话。

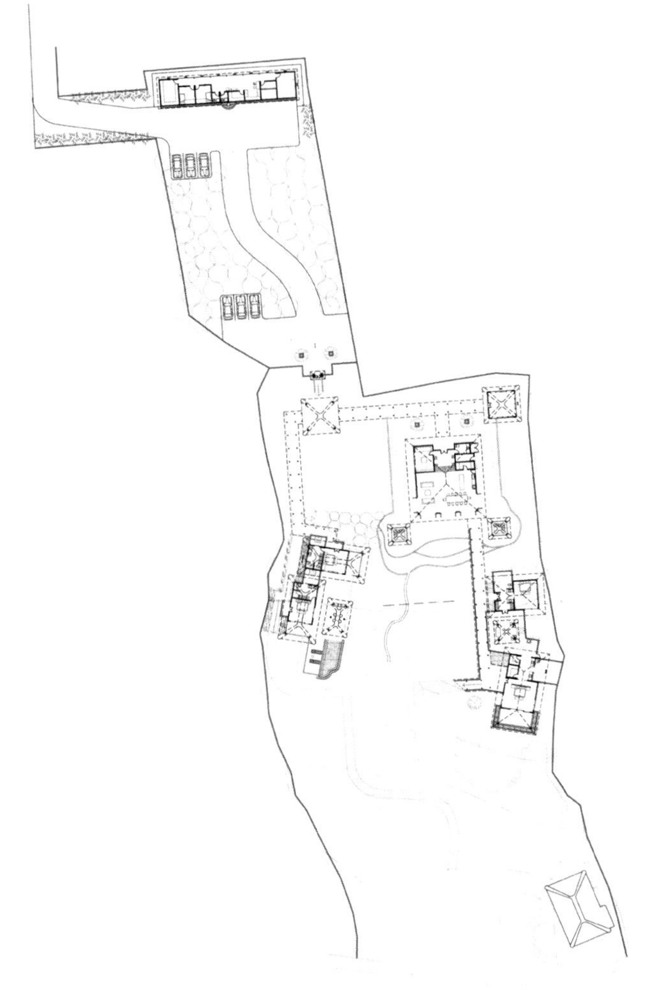

Plan 平面图

Ranch

Ranch 别墅

- Design Agency: Galeazzo Design
- Location: Sao Paulo, Brazil
- Photographer: Maira Acayaba

- 设计单位：Galeazzo 设计机构
- 项目地点：巴西，圣保罗
- 摄影：麦拉・阿卡亚巴

The place was idealized with the basic principles of a private hotel and it is replete of spaces with independent operation, allowing that the many ge¬nerations and guests can enjoy the spaces with privacy. An old elevator was restored and connects all the 3 floors, making the circulation between spaces easier.

In the external area, fountains, lakes, waterfalls and several gardens am¬plifies the fun and the relaxing feeling, such as the rose gardem that was de¬signed for the smokers, the hamick area to enjoy the view and sunset. In the doorway that is a wide pergola made of giant bamboos.

本案是一个拥有许多独立空间的私人酒店，设计的原则是为客人提供私密的休闲空间。一台旧电梯连通 3 个楼层，使各个空间的交通更为便捷。在酒店的外部区域，喷泉、人工湖、瀑布和花园的组合营造了一种轻松愉快的休闲氛围。其中，玫瑰园是专为吸烟的客人而设。露台则供人观景和欣赏日落时分的景色。门前由竹子搭成的藤架，更成为酒店一道亮丽的风景。

Ranch House

乡村小屋

- Design Agency: Galeazzo Design
- Location: Sao Paulo, Brazil
- Area: 1900m^2
- Photographer: Ma í ra Acayaba

- 设计单位：Galeazzo 设计机构
- 项目地点：巴西，圣保罗
- 项目面积：1900m^2
- 摄影：麦拉・阿卡亚巴

Located in the countryside, near to the city of São Paulo, this ranch house of 1900m² was designed to be a place of leisure and meeting to several gene rations of a big family.

For cladding were chosen natural and sustainable materials such as de molition wood to the floor and cabinetry parts, bamboo for the roof overlay and pergolas, rough stone for the fireplace and some walls and soil and calcareous paint. In the main hall an old log was brushed and used as axis for a helical staircase made of corten steel.

The house was divided in sections. Firstly, the spa space includes game room, massage room, steam room, dry sauna, turkish bath, leisure room, gym, heated pool with current and spa for 30 people, both with retractable roof. Secondly, the internal social area includes big living with 4 spaces, dining room, lunch room, social hall, pub, cellar and movie theatre. And then the external social area: big pool with spa, hamick area and a restaurant up to 100 people with infrastructure for bar area, barbacue and pizza, and a little stage for performances. Finally, the intimate area has 7 suites and annex for guests. For the decoration, furniture and lamps with contemporary design, rugs made of straw yields and wool, and vintage pieces.

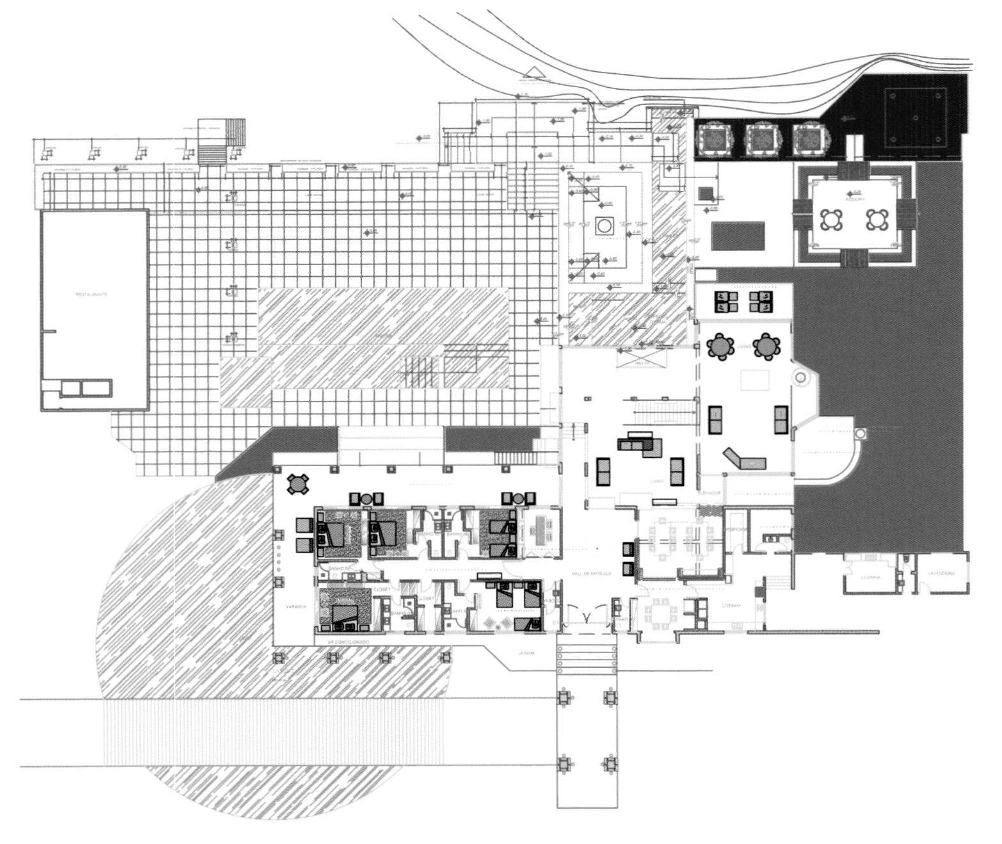

Plan　平面图

本案位于靠近圣保罗的一个乡村，这幢 1900m² 的房子是一个拥有几代人的大家庭用于休闲娱乐和会客的处所。

房子的设计选用了自然环保的材料，比如从地板和木柜拆借的木头，用于屋顶和藤架的竹子，砌壁炉的石头以及涂墙的泥土和石灰涂料。主厅有座柯尔顿钢制的楼梯，围绕一根大木柱子螺旋向上。室内空间被分为几部分。首先，SPA 空间由游戏室、按摩房、湿蒸房、干蒸桑拿房、土耳其浴室、休闲区、健身房和可供 30 人使用的加热游泳池组成，其中健身房和加热游泳池有可灵活移动的顶棚。第二，室内会客交流区包括一个大客厅、主餐厅、午餐厅、会客厅、吧台、酒窖和电影放映厅。而室外会客交流区则包括一个带 SPA 功能的大型游泳池和一个能容纳 100 个人用餐的餐厅，该餐厅拥有吧台、烧烤区和 pizza 区，还有一个用于表演的小舞台。最后，这里还为客人准备了 7 个套房。房间的装饰物、家具和灯具都具有现代风格的设计，由稻草和羊毛制成的地毯，还有复古的小物件，均为室内的设计增添了一丝华彩。

The Parade Ascot Vale

阿斯科特维尔别墅

- Design Agency: Bagnato Architects
- Designer: Marie Bagnato
- Location: Melbourne, Australia
- Area: 262m^2
- Photographer: Axiom Photography

- 设计单位：Bagnato 建筑事务所
- 设计师：玛丽·巴格纳图
- 项目地点：澳大利亚，墨尔本
- 项目面积：262m^2
- 摄影：Axiom 摄影机构

With signature attention to impeccable finish and effortlessly sophisticated function, this home has been transformed into a beautifully luxurious contemporary residence. Throughout the house natural materials have been used such as timber, stone & copper and natural render wall finishes.

Beyond its quietly unassuming facade, intelligent design has created a substantial floor plan that takes full advantage of its unusually deep allotment and highly sought after northerly rear aspect. Three downstairs bedrooms all with built in Robes, the main with a private bathroom and dual Walk in Robes, precede spectacular open plan living and dining zones incorporating a magnificent honed marble kitchen with high end appliances.

Integrated sound and a clever winter garden join built in media storage and an open fire as fitting complements to discerning lifestyle spaces that extend to a night lit landscaped courtyard with tranquil pond, rusted feature wall and built in BBQ for unforgettable alfresco entertaining.

Upstairs an entirely flexible sitting or media room with built desk and cabinetry readily adapts to study or second main bedroom with private bath room. At the rear, a secluded yet generous home office is privately positioned atop a remote double carport via a rear road.

着重强调别墅的精致装修和完善功能，设计师将其打造成了一栋奢华雅致的现代住宅。

设计采用了如木材、石材和铜材等自然材料以及环保的墙漆。该别墅低调朴实的外观之下，设计师充分发挥了空间优势，为这栋面积不算大的别墅提供了一个充实丰富的空间设计。位于一楼的三个房间，拥有独立的衣柜，主卧有一个私人浴室和双门大衣柜。开放式的用餐区和活动区，与厨房相连，厨房中的大理石流理台，光洁而素雅，还配有高端的厨具和设施。景致优美的后院中，清幽的池子和斑驳的墙壁营造出一种与自然和谐亲近的氛围，还为住户提供了可以进行户外烧烤的休闲场地。二楼是一个多用空间，书桌等木艺家具可作为书房用具。这里还有一个卧室，拥有独立的卫浴。另外，一个隐蔽而宽敞的家庭工作室位于二楼后方区域。

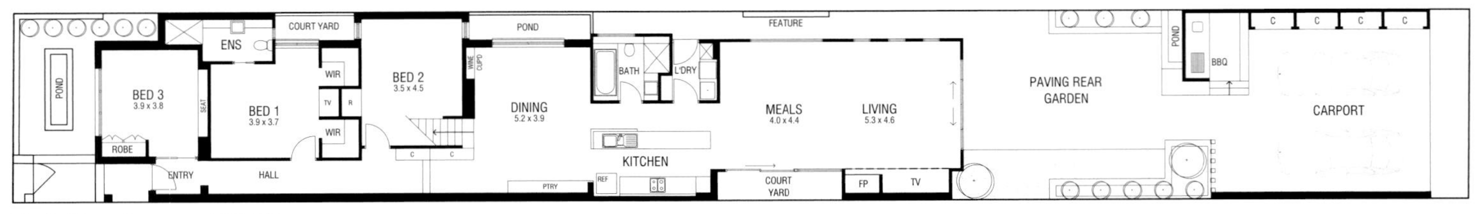

The first floor plan　　一层平面图

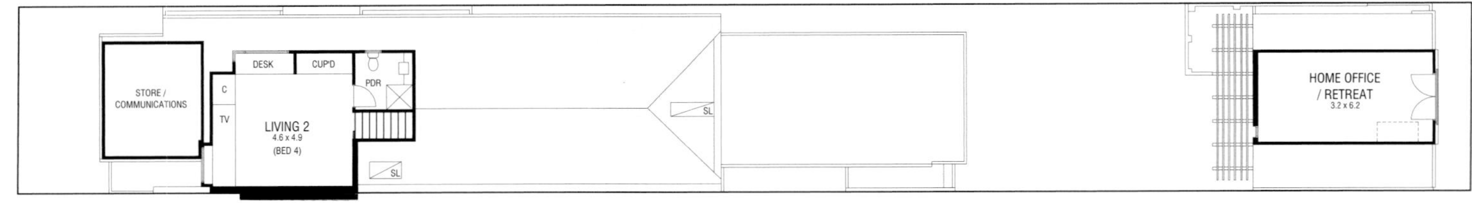

Second floor plan　　二层平面图

The Church Cellar

教堂别墅

- Design Agency: Bagnato Architects
- Designer: Marie Bagnato
- Location: Melbourne, Australia
- Area: 520m^2
- Photographer: Axiom Photography

- 设计单位：Bagnato 建筑事务所
- 设计师：玛丽・巴格纳图
- 项目地点：澳大利亚，墨尔本
- 项目面积：520m^2
- 摄影：Axiom 摄影机构

he site contained both the original church building to one side of the site and on the other side of the site there was an old church multi-purpose hall that was not of any significance. The designers were allowed to demolish the hall and they retained the church structure. The church building was fully renovated with a new roof and where possible they restored the original timber cladding & replaced the timber cladding that was rotten. The 100 year old double hung windows were also dismantled and fully restored to a working condition. At the rear of the church and part of the side we completely opened up the building and installed huge steel beams so that they could install large glazed windows to allow natural light to enter and to have views to the garden. The designers were very mindful to retain as much of the building as possible so that you could see the original shape even though contemporary additions were added.

本案是一栋由一座老教堂改建而成的私人别墅，教堂主楼原始结构和一侧的大厅得到保留。因为教堂是一栋很有年头的老建筑，因此业主在竞拍成功后，按照政府的要求，保留了主楼的原始结构，并对一侧的大厅进行改建。教堂的屋顶被修葺一新，设计师保留了覆盖在屋顶的原始木料，而那些因年代久远而腐化的木材则被替换。老旧的双提拉窗经过拆改修复，焕然一新。教堂的后部和侧边区域被完全打开，加装了钢大梁以装上落地窗，从而让光线自然涌入，使人即使身处室内，也能欣赏到庭院风景。设计师尽其所能保留老教堂的原貌，因此即便是在设计中加入了诸多现代元素，教堂旧时的风貌仍旧依稀可辨。

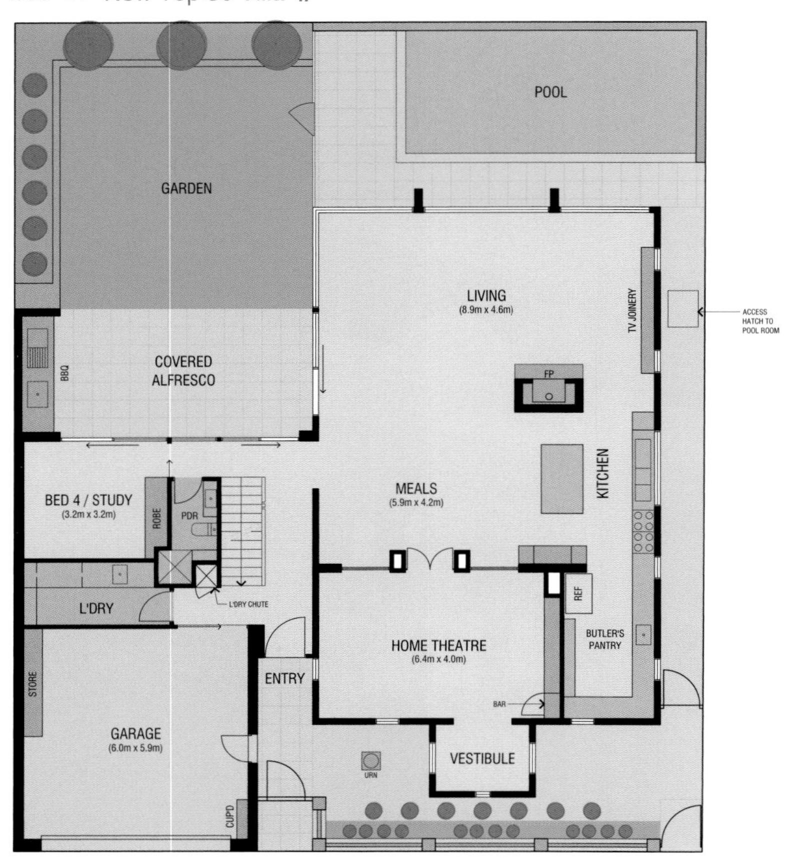
POOL
GARDEN
LIVING
(8.9m x 4.6m)
TV JOINERY
ACCESS HATCH TO POOL ROOM
BBQ
COVERED ALFRESCO
FP
KITCHEN
MEALS
(5.9m x 4.2m)
BED 4 / STUDY
(3.2m x 3.2m)
ROBE
PDR
L'DRY
L'DRY CHUTE
REF
BUTLER'S PANTRY
HOME THEATRE
(6.4m x 4.0m)
ENTRY
STORE
GARAGE
(6.0m x 5.9m)
BAR
VESTIBULE
URN
CUP'D

Plan　　平面图

BEDFORD PK
CONCOURSE
207TH ST
WASH. HTS
168TH ST
6th AVE EXPRESS

Levin Residence

Levin 住宅

- Design Agency: Ibarra Rosano Design Architects
- Location: Marana, Arizona, USA
- Area: $327m^2$
- Photographer: Bill Timmerman

- 设计单位：Ibarra Rosano 建筑设计机构
- 项目地点：美国，亚利桑那州
- 项目面积：$327m^2$
- 摄影：比尔・蒂默曼

Three simple volumes hover above the desert, responding to the challenges of a sloping site and to an ethic of building with minimal disruption to the natural environment. The available buildable area was bifurcated by a minor drainage-way, which inspired the architects to leave the cars behind and link the parking area to the main house by a bridge that allows rainwater and wildlife to flow beneath it. Cantilevered concrete slabs enabled the house to run perpendicular to the topography for optimal solar exposure, for cross-ventilation, and to frame views. Meanwhile, the underside provides shady refuge for desert animals. The tubular forms crop the desert landscape into more intelligible vignettes much like a photographer's square, celebrating the natural setting and view of the city lights the owner did not even realize they had.

在这片荒漠上矗立着三个简单的建筑，整体设计克服了斜面地基带来的挑战，并且实现了对周围自然环境的最小破坏。

由于主要引流系统将这片土地上适合建屋的区域分为了两部分，因此设计师巧妙地将停车区安排在后方与主屋连接处，并用一座小桥连接，使雨水或路过的野生动物能够自桥下而过。外挑的混凝土楼板使房子与地形垂直，以获得最佳的光照和通风条件，以及最好的视野。同时，还为沙漠的动物提供了一个荫蔽之处。设计师利用两头相通的管状空间设计将沙漠风景切成一小片一小片，仿如摄影师精心设置的照片展示区，向人们诉说着连业主自身都未曾发觉的自然环境与城市灯光相互碰撞带来的美景。

图书在版编目（CIP）数据

最新别墅设计50例Ⅱ/深圳市海阅通文化传播有限公司主编.—北京：中国建筑工业出版社，2015.10
ISBN 978-7-112-18533-7

Ⅰ.①最… Ⅱ.①深… Ⅲ.①别墅-室内装饰设计-世界-图集 Ⅳ.①TU241.1-64

中国版本图书馆CIP数据核字(2015)第234778号

责任编辑：费海玲 张幼平 王雁宾
责任校对：李美娜 姜小莲

最新别墅设计50例Ⅱ
深圳市海阅通文化传播有限公司 主编
*
中国建筑工业出版社出版、发行（北京西郊百万庄）
各地新华书店、建筑书店经销
深圳市海阅通文化传播有限公司制版
北京缤索印刷有限公司印刷
*
开本：889×1194毫米 1/20 印张：10⅗ 字数：300千字
2016年2月第一版 2016年2月第一次印刷
定价：98.00元
ISBN 978-7-112-18533-7
(27664)
版权所有 翻印必究
如有印装质量问题，可寄本社退换
（邮政编码 100037）